F. Link

Eclipse Phenomena
in Astronomy

With 158 Figures

Springer-Verlag New York Inc. 1969

Professor Dr. F. Link
Institut d'Astrophysique, Paris

All rights reserved. No part of this book may be translated or reproduced in any form without written permission from Springer-Verlag. © by Springer-Verlag Berlin Heidelberg 1969 · Library of Congress Catalog Card Number 68-56208. Printed in Germany

The use of general descriptive names, trade names, trade marks, etc. in this publication, even if the former are not especially identified, is not to be taken as a sign that such names, as understood by the Trade Marks and Merchandise Marks Act, may accordingly be used freely by anyone

Title No. 1541

To the memory of

André Danjon
(1890–1967)

Preface

Eclipses and problems related to them have been, from ancient times, one of the main interests not only of astronomers but indeed of all mankind. The appearance of eclipses, lunar as well as solar, excited the imagination of our ancestors and provoked their curiosity to explain their origin or to use them for the further investigation of celestial bodies.

With the present development of astronomy the eclipse problems are not limited to the Sun and the Moon, as in the past, but have been progressively extended to the components of the solar system and to domains of radiations other than optical ones. Our intention is to give an account of all these problems in their theoretical and experimental form with some additions on their historical development. Those of our readers not interested in the historical side may feel at first inclined to ignore this part, but we are sure that eventually they will be sufficiently interested to repair this omission.

It would be possible to construct our work on a common and vastly general frame of mathematical formulae and then treat each problem as a particular solution of the general formulae. This approach may be perhaps the most economical but according to our experience is not the best. However, a more readable way will be the independent treatment of different problems with some references on the similarity or analogy to others. In this way we leave the reader or student to detect for himself the general framework of eclipse phenomena as this will be a very good test of the comprehension of our work. A further advantage of this presentation is that some readers who are interested in only one of the problems need not to study the general and necessarily abstract formulae.

Our work is divided into 7 principal sections as follows:
1. Lunar eclipses
2. Eclipses of artificial Earth satellites
3. Twilight phenomena
4. Occultation and eclipses by other planets
5. Transits of planets over the Sun
6. Eclipse phenomena in radio astronomy
7. Einstein's deflection of light.

It appears from this enumeration that stress has been laid on the part played by the eclipsing body. For this reason solar eclipses have not been included in this work. As for eclipsing variables, the matter of which is generally far from the methods used in the above phenomena, there are already several good monographs on this subject.

An invitation to give the semestrial lecture at the Faculté des Sciences de Paris (1967/68) was the origin of the following text which was, in addition, accepted by Springer Publishing House. My thanks are, therefore, due equally to both.

Paris, February 1968   F. LINK

Contents

1. Lunar Eclipses

1.1.	Introduction to Lunar Eclipses	1
1.1.1.	Geometrical Conditions of Lunar Eclipses	1
1.1.2.	Computation of a Lunar Eclipse	3
1.1.3.	Geographical Circumstances of Eclipses	5
1.1.4.	Future Lunar Eclipses	8
1.1.5.	The History of Lunar Eclipses	9
1.1.6.	Lunar Eclipses in Chronology	11
Bibliography		12
1.2.	Photometrical Theory of the Umbra	13
1.2.1.	Eclipse Scene	13
1.2.2.	General Photometrical Theory of the Umbra	14
1.2.3.	Integration in the Solar Plane	15
1.2.4.	General Transmission Coefficient	19
1.2.5.	Molecular Scattering of Light	20
1.2.6.	Aerosol Scattering	21
1.2.7.	Astronomical Determinations	21
1.2.8.	Terrestrial Method	22
1.2.9.	Discussion of A	24
1.2.10.	Attenuation of Light by Refraction	24
1.2.11.	Auxiliary Shadow	27
1.2.12.	The Normal Densities of the Shadow	28
1.2.13.	Theory of Refraction and Air Mass	33
1.2.14.	Confrontation of the Refraction Theory with Observations	35
1.2.15.	Climatic Influences on the Refraction and Air Mass	36
1.2.16.	Climatic Variations of the Shadow Density	38
1.2.17.	High Absorbing Layers	40
1.2.18.	Atmospheric Illumination of the Eclipsed Moon	42
1.2.19.	Eclipse Phenomena in the Cislunar Space	46
1.2.20.	Lunar Eclipses on the Moon	47
1.2.21.	Other Photometrical Theories	49
1.2.22.	Old and Classical Theories of Refraction and Air Mass	51
Bibliography		54
1.3.	Photometry of Lunar Eclipses	55
1.3.1.	Measurements of the Shadow Density	55
1.3.2.	Visual Method	55
1.3.3.	Photographic Method	57
1.3.4.	Photoelectric Method	58
1.3.5.	Comparison of the Theory with Observations	58

1.3.6.	Atmospheric Ozone	62
1.3.7.	High Absorbing Layer	66
1.3.8.	The Behaviour of Meteoritic Particles in the Atmosphere	71
1.3.9.	Tropospheric Influences	77
1.3.10.	Meteorological Perturbations	78
1.3.11.	Atmospheric Pollution of Planetary Extent	80
1.3.12.	Global Intensity of the Eclipsed Moon	84
1.3.13.	Surveyor III Eclipse Observation from the Moon	85
Bibliography		87
1.4.	Lunar Luminescence	88
1.4.1.	Simplified Theory of the Penumbra	88
1.4.2.	Complete Theory of the Penumbra	89
1.4.3.	Comparison with Observations	91
1.4.4.	Interpretation of the Light Excess	91
1.4.5.	Lunar Luminescence	92
1.4.6.	Fluctuations of the Global Luminosity of the Moon	94
1.4.7.	Recent Work on Lunar Luminescence	96
1.4.8.	Brightness of Lunar Eclipses and Solar Activity	97
1.4.9.	Danjon's Relation and the Solar Cycle	100
1.4.10.	Problematic Variations of the Penumbral Density	102
Bibliography		103
1.5.	Increase of the Shadow	104
1.5.1.	Short History of the Shadow Increase	104
1.5.2.	Maedler's Method	105
1.5.3.	Hartmann's Method	106
1.5.4.	Kosik's Method	106
1.5.5.	Results on the Shadow Increase	108
1.5.6.	Explanation of the Shadow Increase	110
1.5.7.	Paetzold's Experiments	111
1.5.8.	Connection between the Shadow Increase and the Meteoric Activity	112
1.5.9.	Explanation of the Shadow Flattening	113
Bibliography		114
1.6.	Thermal Phenomena during Lunar Eclipses	115
1.6.1.	Brief Outline of Temperature Measurements	115
1.6.2.	Theoretical Aspects of Surface Temperature Variations	116
1.6.3.	Methods of Observing the Thermal Radiation of the Moon	117
1.6.4.	Experimental Results	118
1.6.5.	Hot Spots on the Moon	119
1.6.6.	Lunar Eclipse at Microwaves	120
Bibliography		121

2. Eclipses of Artificial Earth Satellites

2.1.	Preliminary Remarks	122
2.2.	General Conditions of Visibility	123
2.3.	Ephemeris of the Eclipse	124
2.4.	Photometrical Theory of the Shadow	128
2.5.	General Transmission Coefficient	129

2.6.	Simplified Presentation of the Eclipse Theory	131
2.7.	Eclipse Observed from the Satellite	141
2.8.	Secondary Illuminations	143
2.9.	Fesenkov's Treatment of the Eclipse Problem	144
2.10.	Observing Methods Used for Passive Satellites	146
2.11.	Work at the Naval Ordnance Test Station, China Lake, California	147
2.12.	Work at Ondřejov Observatory on Echo-2 Eclipses	150
2.13.	Work at Valensole Station	151
2.14.	Active Satellites of SR Series	152
2.15.	Eclipses of the Satellite SR-I	154
2.16.	Work of Arcetri Group	157
2.17.	Work of Slough Group	160
Bibliography		168

3. Twilight Phenomena

3.1.	Different Components of the Twilight	169
3.2.	Night Light Correction	170
3.3.	Illumination of the Upper Atmosphere	171
3.4.	Fundamental Problems of Twilight Phenomena	174
Bibliography		179

4. Occultations and Eclipses by Other Planets

4.1.	General Remarks	180
4.2.	Dioptrics of the Thin Planetary Atmosphere	180
4.3.	Planetary Atmosphere from the Refraction Standpoint	182
4.4.	Basic Equations for the Far Occultation	184
4.5.	Theoretical Course of a Far Occultation	186
4.6.	Occultation of Regulus by Venus, July 7, 1959	187
4.7.	Occultation of σ Arietis by Jupiter, November 20, 1952	189
4.8.	Near Occultation by the Planet	192
4.9.	Eclipses of Phobos	193
4.10.	Reduction of Eclipse Curves of Jovian Satellites	197
4.11.	Eropkin's Eclipse Curves of Jovian Satellites	200
4.12.	Frost Phenomena on Jovian Satellites during Their Eclipses	202
4.13.	Terrestrial Occultation Observed from the Moon	203
Bibliography		204

5. Transits of Planets over the Sun

5.1.	Introductory Remarks	205
5.2.	Refraction in the Planetary Atmosphere	205
5.3.	Refraction Image of the Sun	206
5.4.	Discussion of Past Transits	209
5.5.	Simplified Discussion of the Transit Phenomena	211

5.6.	Lomonosov's Phenomenon	215
5.7.	Extension of the Cusps of Venus	216
5.8.	Explanations of the Cusps Extension	218
5.9.	Edson's Work on Cusps Extension	221
5.10.	New Investigations on Cusps Extension	222
Bibliography		225

6. Eclipse Phenomena in Radio Astronomy

6.1.	Introduction	226
6.2.	Occultation Scene	227
6.3.	Occultations of Radio Sources by the Moon	228
6.4.	Numerical Example	229
6.5.	Occultation of Radio Source by Solar Corona	231
6.6.	Modification of Light Intensity	232
6.7.	Observations of Coronal Occultations	234
6.8.	Occultation of the Mariner-IV Space Probe by Mars	236
6.9.	Further Implications of the Mariner-IV Mission	240
6.10.	Occultation of the Mariner-V Space Probe by Venus	241
Bibliography		243

7. Einstein's Deflection of Light

7.1.	Einstein's Deflection of Light	244
7.2.	Dioptrics of Einstein's Deflection	246
7.3.	Photometry of Einstein's Deflection	248
7.4.	Other Expression of Illumination	250
7.5.	Illumination by Stellar Disk	253
7.6.	Normalised Gravitational Occultation	255
7.7.	Images of the Occulted Star	257
7.8.	Tichov's Investigations	260
7.9.	Refsdal's Investigation	262
7.10.	Liebes' Investigations	262
7.11.	Consequences of Einstein's Deflection in the Stellar Universe	263
7.12.	Gravitational Passage or Occultation	265
7.13.	History of Einstein's Photometrical Effect	267
Bibliography		268
Subject Index		269

1. Lunar Eclipses

1.1. Introduction to Lunar Eclipses

1.1.1. Geometrical Conditions of Lunar Eclipses

The geometrical conditions of a lunar eclipse are given by the two cones formed through the common tangents interior and exterior to the Sun and the Earth (Fig. 1.1.1). At the distance of the Moon these cones give with the plane normal to the axis Sun–Earth two circular intersections named the *penumbra* and the *umbra*.

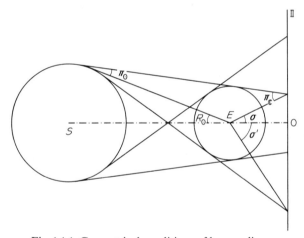

Fig. 1.1.1. Geometrical conditions of lunar eclipses

Their angular semidiameters as seen from the center of the Earth are

$$\sigma = \pi_{\mathrm{C}} + \pi_{\odot} - R_{\odot} \quad \text{for the umbra}$$

and

$$\sigma' = \pi_{\mathrm{C}} + \pi_{\odot} + R_{\odot} \quad \text{for the penumbra,} \tag{1.1}$$

are $\pi_{\odot}$ and π_{C} where the parallaxes of the Sun and the Moon and $R_{\odot}$ is the solar angular radius.

[1] Link, Eclipse Phenomena

When the Moon passes near to the shadow axis *SE* a lunar eclipse of one of the following forms may occur (Fig. 1.1.2):

a) *Penumbral eclipse* when the Moon penetrates only into the penumbra.

b) *Partial eclipse* when the Moon penetrates also partially into the umbra.

c) *Total eclipse* when the Moon enters wholly into the umbra. A special case may arise if the Moon's path goes through the center of the umbra, but such a central eclipse is extremely rare.

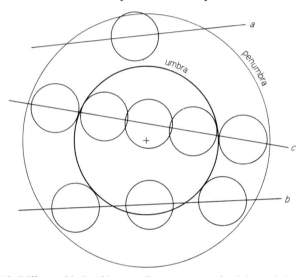

Fig. 1.1.2. Different kinds of lunar eclipses: a penumbral, b partial, c total

It is now realistic to place oneself in the position of a lunar observer. From this point of view the partial eclipse of the Sun takes place at all lunar spots situated in the penumbra and the total eclipse in the umbra.

In connection with the geometrical conditions of lunar eclipses we shall give here a brief account of the phenomenon as it appears to the ordinary observer: The entry into the penumbra is not perceptible until about half the lunar diameter is covered by it. At this moment we can detect a small degradated obscuration which becomes more and more pronounced as the lunar border approaches the umbra. The entry into it is more or less definite, and even an untrained observer can determine its moment with an error of less than $\frac{1}{2}$ minute.

At the beginning of the partial phase the part of the disk obscured by the umbra is nearly invisible due to the contrast with the bright crescent in the penumbra. When a considerable part of the lunar disk has entered into the umbra we generally begin to distinguish the obscured part in contours and later in some details such as maria and bright craters. The total eclipse may be coloured and bright or grey and dark, as will be described later. The gradation of the luminance over the disk sometimes gives the illusion of a sphere. During the totality the sky becomes as dark as during a moonless night and small stars may be seen in the vicinity of the lunar disk. The course of the phenomena during the second half of the eclipse is usually symmetrical but some deviation in colour or brightness may be observed.

In any case, the total eclipse of the Moon is a very picturesque astronomical phenomenon and merits the attention of everybody, not only astronomers who are usually too busy gathering scientific data to appreciate its beauty.

1.1.2. Computation of a Lunar Eclipse

This is one of simplest problems in spherical astronomy especially when we adjust the accuracy of computations to that of observations. We assimilate the part of celestial sphere where the eclipse should be observed to the tangential plane of the sphere at the antisolar point, i.e. at the intersection of the shadow axis with the celestial sphere. The astronomical ephemeris gives us the following elements:

T — the moment in G.M.T. of the opposition in the right ascension of the Moon and the Sun,

$\alpha_\odot, \alpha_\mathrm{C}$ — the corresponding right ascensions of the Sun and the Moon,

$\delta_\odot, \delta_\mathrm{C}$ — the declinations of both bodies,

$\left.\begin{array}{l}\Delta\alpha_\odot, \Delta\alpha_\mathrm{C} \\ \Delta\delta_\odot, \Delta\delta_\mathrm{C}\end{array}\right\}$ the hourly motions of the Sun and the Moon,

$\left.\begin{array}{l}\pi_\odot, \pi_\mathrm{C} \\ R_\odot, R_\mathrm{C}\end{array}\right\}$ the parallaxes and the angular semidiameters.

The center of the shadow O (Fig. 1.1.3) on the sphere has the coordinates $\alpha_\odot + 180°$, $-\delta_\odot$ with their hourly variations $\Delta\alpha_\odot$, $-\Delta\delta_\odot$.

We shall now trace the path of the Moon projected into our plane relative to the center O in a rectangular coordinate system whose $+Oy$ axis is directed towars the north celestial pole and whose $+Ox$ axis has the direction of the daily motion i.e. towards the west. At the instant of the opposition the coordinates of the Moon's center are

$$X_0 = 0$$
$$Y_0 = \delta_\mathrm{C} + \delta_\odot \qquad (1.1a)$$

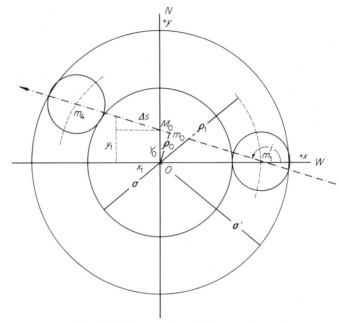

Fig. 1.1.3. Computation of a lunar eclipse

and one hour later

$$x_1 = -15 \cos \delta_\odot (\Delta\alpha_\mathrm{C} - \Delta\alpha_\odot)$$
$$y_1 = \delta_\mathrm{C} + \delta_\odot + \Delta\delta_\mathrm{C} + \Delta\delta_\odot. \tag{1.1b}$$

The hourly path of the Moon has also the length

$$\Delta s = +\sqrt{x_1^2 + (y_1 - Y_0)^2} = -\frac{\Delta\delta_\mathrm{C} + \Delta\delta_\odot}{\sin i} = \frac{15 \cos \delta_\odot (\Delta\alpha_\mathrm{C} - \Delta\alpha_\odot)}{\cos i}, \tag{1.1c}$$

and the inclination towards the $+Ox$ axis

$$\mathrm{tg}\, i = -\frac{\Delta\delta_\mathrm{C} + \Delta\delta_\odot}{15 \cos \delta_\odot (\Delta\alpha_\mathrm{C} - \Delta\alpha_\odot)}. \tag{1.1d}$$

At the middle of the eclipse the minimum distance between the Moon's and shadow centers is

$$\rho_0 = (\delta_\mathrm{C} + \delta_\odot) \cos i \tag{1.1e}$$

and the path

$$\overline{m_0 M_0} = (\delta_\mathrm{C} + \delta_\odot) \sin i$$

so that the moment of the middle will be

$$t_0 = T - \frac{\delta_\mathrm{C} + \delta_\odot}{\Delta\delta_\mathrm{C} + \Delta\delta_\odot} \sin^2 i. \tag{1.2}$$

Introduction to Lunar Eclipses

The beginning of the partial phase or its end take place at the moment when the distance of the two centers Moon-antisun reach the value $\rho_1 = \sigma + R_{\mathrm{C}}$ and the paths to be covered are

$$\overline{m_0 m_1} = \overline{m_0 m_4} = +\sqrt{\rho_1^2 - \rho_0^2},$$

so that the corresponding moments are

$$t_{1,4} = t_0 \pm \frac{\sqrt{\rho_1^2 - \rho_0^2}}{\Delta\delta_{\mathrm{C}} + \Delta\delta_{\odot}} \sin i. \qquad (1.3)$$

For the total phase we get in the analogous way with $\rho_2 = \sigma - R_{\mathrm{C}}$

$$t_{2,3} = t_0 \pm \frac{\sqrt{\rho_2^2 - \rho_0^2}}{\Delta\delta_{\mathrm{C}} + \Delta\delta_{\odot}} \sin i \qquad (1.4)$$

and finally for the external contacts with the penumbra $\rho_1' = \sigma' + R_{\mathrm{C}}$

$$t'_{1,4} = t_0 \pm \frac{\sqrt{\rho_1'^2 - \rho_0^2}}{\Delta\delta_{\mathrm{C}} + \Delta\delta_{\odot}} \sin i. \qquad (1.5)$$

The magnitude of the eclipse is given by the ratio

$$g = \frac{\sigma - \rho_0 + R_{\mathrm{C}}}{2R_{\mathrm{C}}} \qquad (1.6)$$

which is the fraction of the eclipsed lunar diameter at the middle of the phenomenon. In the past the magnitude was usually expressed by "inches" according the rule: 12 inches $= 2R_{\mathrm{C}}$. We have therefore

$g > 1 = 12$ inch   for the total eclipse

$0 < g < 1$ for the partial eclipse.

In the above formulae the motions of the Moon and of the Sun were assumed to be uniform. This approximation holds good for ordinary prediction giving an accuracy within several seconds of time.

Obviously our formulae may be used for the purpose of graphic representation of the eclipse. Drawn with care and on a suitable scale, the diagram of the eclipse can supersede the computations if we content ourself with an accuracy of ± 1 min.

1.1.3. Geographical Circumstances of Eclipses

The limit of the shadow on the Moon corresponds in some geometrical way to the terminator of the shadow on the Earth's surface. It is not without interest to establish the necessary mathematical relations [1] which may be of use for detailed discussion of observations as the density of the shadow depends on local structure of the atmosphere variable with the latitude.

According to the astronomical definition used in standard ephemeris the terminator of the shadow will be assumed here as the line on the Earth's surface where (simultaneously) the upper limb of the Sun rises or sets at the given moment. Thus its angular distance from the subsolar point S' (Fig. 1.1.4) is $90° + (\omega/2) + R_\odot$ or from the antisolar point $90° - (\omega/2) - R_\odot$ where ω is the twofold astronomical refraction on the horizon. To fix the position of the terminator in the chart we can use the tables given in the Nautical Almanac or American Ephemeris where for each 10th day the rising or the setting of the Sun is given as a function of the latitude. These times indicate directly the east longitude of the

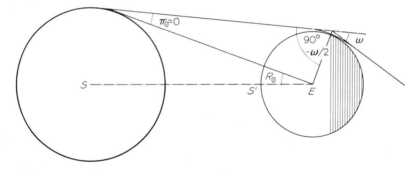

Fig. 1.1.4. Formation of the shadow terminator

terminator on the corresponding latitude for the midnight of U.T. At any other moment H of U.T. the terminator will be shifted westward by H.

Generally the whole terminator does not take part in the formation of the shadow but only its arc $S_1 S_2$, given by the central angle ΔP on both sides of its mid point S_0 (Fig. 1.1.5). Let us consider the geometrical aspect of the eclipse from the lunar point N given by its angular distance γ from the center of the shadow and the position angle P. For the lunar observer at N the Sun subtends the angle

$$R'_\odot = \frac{\pi_{\mathbb{C}}}{\pi_{\mathbb{C}} + \pi_\odot} R_\odot \approx 0.996 \, R_\odot$$

and its angular distance from the Earth's center is

$$\gamma' = \frac{\pi_{\mathbb{C}}}{\pi_{\mathbb{C}} + \pi_\odot} \gamma \approx 0.996 \, \gamma.$$

Thus we have for
$$\sin \Delta P = \frac{R'_\odot}{\gamma'} = \frac{R_\odot}{\gamma}. \tag{1.7}$$

From the spherical triangle $S_0 P_N E'$ we find

$$\cos P = \frac{\sin \varphi}{\sin\left[90° - \frac{\omega}{2} - R_\odot\right]\cos \delta_\odot} - \operatorname{ctg}\left[90° - \frac{\omega}{2} - R_\odot\right]\operatorname{tg} \delta_\odot$$

or with a good approximation
$$\cos P = \frac{\sin \varphi}{\cos \delta_\odot}. \tag{1.8}$$

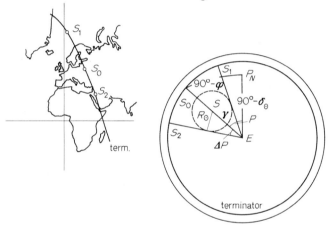

Fig. 1.1.5. Geographical conditions of a lunar eclipse

Therefore, the end points of the effective part of the terminator will have the following latitudes

$$\sin \varphi_{1,2} = \cos \delta_\odot \cos(P \mp \Delta P). \tag{1.9}$$

From the beginning of the penumbral eclipse where $\Delta P = 0$, the length of the effective terminator increases and attains its greatest value at the distance $R_\odot$ where the whole terrestrial circumference of 40,000 km participates in the formation of the shadow. Thus in external parts of the umbra the local structure of terrestrial atmosphere can play some role in contrast with central regions, where the average atmospheric structure is preponderant.

The numerical values of φ, λ and P for the eclipse January 19, 1954 are given in the Table 1.1. Its graphical representation will be given later (Fig. 1.3.6).

Table 1.1. *Shadow terminator 1954 I 19 at 2ʰ GMT*

φ	−69,5°	−60°	−40°	−20°	0°	+20°	+40°	+60°	+70,5°
					Sunrise				
λ	27°W	20°E	43°E	54°E	62°E	70°E	80°E	101°E	153°E
P	180°	158°	133°	111°	90°	69°	47°	22°	0°
					Sunset				
λ	27°W	75°W	98°W	109°W	116°W	124°W	134°W	155°W	153°E
P	−180°	−158°	−133°	−111°	−90°	−69°	−47°	−22°	0°

1.1.4. Future Lunar Eclipses

Table 1.2. *Umbral eclipses according to* OPPOLZER [2]

(1) The number of the eclipse according to OPPOLZER; (2) the civil date; (3) the Julian day; (4) the middle of the eclipse in G.M.T.; (5) the magnitude of the eclipses in inches (12 inches = the Moon's diameter); (6) the semiduration of the partial phase; (7) the semiduration of the total phase; (8) the geographical longitude; and (9) the geographical latitude of the sublunar point at the time of the conjunction.

(1)	(2)	(3)	(4)		(5)	(6)	(7)	(8)	(9)
			(h)	(min)		(min)		(°)	(°)
4912	1968 IV 13	2439 960	4	49	13.6	103	28	− 72	− 8
4913	1968 X 6	2440 136	11	41	14.1	104	31	−178	+ 5
4914	1970 II 21	2440 639	8	31	0.6	26	—	−124	+11
4915	1970 VIII 17	2440 816	3	25	5.0	71	—	− 50	−14
4916	1971 II 10	2440 993	7	42	15.6	107	39	−112	+14
4917	1971 VIII 6	2441 170	19	44	20.7	112	51	+ 65	−17
4918	1972 I 30	2441 347	10	53	12.9	102	21	−160	+18
4919	1972 VII 26	2441 525	7	18	6.9	80	—	−108	−20
4920	1973 XII 10	2442 027	1	48	1.2	36	—	− 29	+23
4921	1974 VI 4	2442 203	22	14	9.9	93	—	+ 26	−22
4922	1974 XI 29	2442 381	15	16	15.5	106	38	+128	+21
4923	1975 V 25	2442 558	5	46	17.5	109	45	− 87	−21
4924	1975 XI 18	2442 735	22	24	13.1	102	23	+ 20	+19
4925	1976 V 13	2442 912	19	50	1.7	43	—	+ 62	−18
4926	1977 IV 4	2443 238	4	21	2.5	51	—	− 64	− 6
4927	1978 III 24	2443 592	16	25	17.5	109	45	+115	− 2
4928	1978 IX 16	2443 768	19	3	16.0	107	41	+ 73	− 3
4929	1979 III 13	2443 946	21	10	10.5	94	—	+ 45	+ 3
4930	1979 IX 6	2444 123	10	54	13.4	103	26	−164	− 7
4931	1981 VII 17	2444 803	4	48	6.9	80	—	− 71	−21
4932	1982 I 9	2444 979	19	56	16.2	107	42	+ 63	+22
4933	1982 VII 6	2445 157	7	30	20.6	112	51	−112	−23
4934	1982 XII 30	2445 334	11	26	14.4	105	33	−171	+23
4935	1983 VI 25	2445 511	8	25	4.1	65	—	−126	−23

Table 1.3. *Penumbral eclipses according to* LIU BAO-LIN [3]

(1) The number according to LIU BAO-LIN; (2) the civil date; (3) the Julian day; (4) the fraction of lunar diameter eclipsed by the penumbra; (5 — 7) the beginning, the middle and the end of the eclipse in G.M.T.

(1)	(2)			(3)	(4)	(5)		(6)		(7)	
						(h)	(min)	(h)	(min)	(h)	(min)
4	1969	IV	2	244 0314	0.728	16	39	18	33	20	27
5	1969	VIII	27	0461	0.039	10	22	10	49	11	16
6	1969	IX	25	0490	0.928	18	06	20	10	22	15
7	1973	I	18	1701	0.891	19	17	21	18	23	18
8	1973	VI	15	1849	0.494	19	05	20	51	22	36
9	1973	VII	15	1879	0.130	10	44	11	39	12	34
10	1976	XI	6	3089	0.863	20	47	23	03	01	18
11	1977	IX	27	3414	0.926	06	18	08	29	10	40
12	1980	III	1	4300	0.680	18	44	20	46	22	48
13	1980	VII	27	4448	0.281	17	57	19	10	20	22
14	1980	VIII	26	4478	0.734	01	42	03	32	05	21
15	1981	I	20	4625	1.040	05	36	07	50	10	04
16	1983	XII	19	5688	0.915	23	46	01	49	03	52

1.1.5. The History of Lunar Eclipses

Lunar eclipses are one of the oldest celestial phenomena, reported in historical sources as early as 2283 B.C. [4], the eclipse related to the Mesopotamian town of Ur, or again the Chinese eclipse [5] in 1136 B.C. From the beginning of the 8th century B.C. the number of eclipses observed in Mesopotamia and in the Mediterranean region has been continuously growing with later additions from the rest of Europe. Ancient observed lunar eclipses may be found in the lists assembled by CALVISIUS [6], RICCIOLI [7] and STRUYCK [8]. The Chinese eclipses were collected in the list by GAUBIL [5].

After the foundation of different Academies in the second half of the 17th century, the observed eclipses were reported in their transactions. Also the bibliography by REUSS [9] and for the 17th century by PINGRÉ-BIGOURDAN [10] are of great use.

Since the 19th century observation appeared in different publications of astronomical observatories and in the new astronomical periodicals and other scientific magazines.

The computed lunar eclipses are given in catalogues by OPPOLZER [2], GINZEL [11] and, for penumbral eclipses, by BAO-LIN [3]. As the basic elements for these catalogues are not the same, their indications may differ to some extent.

The part played by lunar eclipses in the development of astronomical knowledge is not negligible [12]. In the 4th century B.C. ARISTOTE [13] saw in the circular form of the shadow edge in every eclipse proof of the spherical shape of the Earth. ARISTARCHOS of Samos in the 3rd century B.C. [14] and HIPPARCHOS in the 2nd century B.C. proposed the use of the lunar eclipses for the determination of the relative dimensions of the system Sun-Earth-Moon. HIPPARCHOS also proposed the first method for the determination of geographical longitudes by simultaneous observations of a lunar eclipse from two distant places [15]. PTOLEMY [16] in the 2nd century A.D., and after him again and again many astronomers right up to the present time, used the old eclipses for investigations or improvements of the very complicated theory of lunar motions.

In the 17th and partly also in 18th century Hipparchos's old method for the determination of longitudes was renovated using the transits of craters on the edge of the shadow as suggested by LANGRENUS [17]. Though the accuracy of this method could not exceed more than some tenth of a minute of time, its utility was great in those times. For instance the eclipse in 1634 observed in Cairo, Aleppo and the western part of Europe, enabled the astronomers to shorten the Mediterranean Sea by 1000 km in respect to its assumed length before that time [18].

Besides the astrometrical aspects of lunar eclipses, some astrophysical considerations appeared from the 17th century, when the visibility of the Moon in umbra intrigued the astronomers. LICETUS expressed [19] the opinion that the visibility of the eclipsed Moon is due to the phosphorescence of lunar material after its long exposure to solar radiations, but this explanation was adopted by only a few other astronomers. Kepler's explanation [20] by the refraction of solar rays in terrestrial atmosphere prevailed soon after and was universally accepted until recent times.

The first photometrical theory dealing with the illumination of the eclipsed Moon was carried out by D. DU SÉJOUR whose work, published under the general title "Traité analytique des mouvements apparents des corps célestes" where lunar eclipses form a small part, remained totally ignored by all subsequent astronomers [21].

At the end of the 19th century interest was raised by the increase of the Earth's shadow especially by its theoretical explanation. The phenomenon, of course, experimentally known from the beginning of the 18th century, was theoretically investigated two centuries later by J. HEPPERGER and H. SEELIGER, while A. BROSSINKY and J. HARTMANN carried out a new reduction of old series of observations for the experimental determination of the increase (1.5). Unfortunately both theories failed to give the expected explanation for reasons given elsewhere.

The last revival brought up the discovery of Danjon's law relating the luminosity of eclipses to solar activity (1.4.9) together with the possibility of exploring the upper atmosphere by means of lunar eclipses. However, that is no longer the true story of eclipses but the theme of our book.

1.1.6. Lunar Eclipses in Chronology

Besides the contribution of lunar eclipses to astronomy we may consider also the important part they playd in chronology. Eclipses were often reported in early historical sources as "epitheton ornans et constans" of many great events. According to the superstitions of the ancients, lunar eclipses were considered as evil omens, due no doubt to the bloody color of the totaly eclipsed Moon. In this way some historical events, especially battles, were directly influenced by the appearance of a lunar eclipse.

These circumstance made lunar eclipses an important aid in chronological research and they have frequently been used, of course, with a thorough analysis of the historical source. This is necessary because many reports are incomplete from an astronomical viewpoint.

Table 1.4. *Some historical eclipses*

Date			Description	Source
−522	VII	16	In the 7th year of the reign of Cambyses one hour before midnight in Babylon the Moon was eclipsed from the North over half of its diameter	PTOLEMY, Syntaxis, V, 14
−412	VIII	27	This eclipse retarded the retreat of the Athenian army under Nicias from Sicily and caused its defeat by the Syracusians	PLUTARCH in Life of Nicias
−336	VIII	19	This eclipse, explained by the clever prophet Miltas of Dion's army, decided its departure for Sicily in order to overthrow the local tyrant Dionisius	PLUTARCH in Life of Dion
−330	IX	20	This eclipse happened eleven days before the battle of Arbela with the victory of Alexander over Darius	PLUTARCH in Life of Alexander
−218	IX	1	Gaulish mercenary troops were greatly alarmed by this eclipse so that Attalus, king of Pergamos, had to get rid of them	POLYBIUS, Histories V, 78

Table 1.4 (Continued)

Date			Description	Source
−171	IX	3	On the eve of the battle of Pydna the eclipse predicted by the Roman tribune, C. Sulpicius Gallus, took place	Livy, Hist. Rom. XLIV, 37
+33	IV	3	This eclipse is generally accepted as the eclipse of the Crucifixion of our Lord	
72	II	22	An example of the horizontal eclipse where the rising Sun and setting eclipsed Moon were visible simultaneously	PLINY, Nat. Hist. II, 3
1504	III	1	Columbus knowing in advance that this eclipse should happen, gained the reputation of a prophet among the Indians who, in consequence, supplied provisions to Spaniard expedition	HELPS, Life of Columbus

Bibliography

1. LINK, F.: Bull. astron. **13**, 3 (1947).
2. OPPOLZER, TH.: Denkschr. Akad. Wiss. Wien, Math.-Phys. Kl. **52** (1887).
3. BAO-LIN, LIU: Tien Wen Hsüe Pao (Acta Astr. Sinica) **12**, No. 1 (1964).
4. SCHOCH, C.: Erg.-Hefte AN **8** (1930).
5. GAUBIL, A.: Observations astronomiques. Paris 1740, V. 3.
6. CALVISIUS, S.: Opus chronologicum. Francoforti 1650.
7. RICCIOLI, J.B.: Almagestum novum. Bononiei 1661.
8. STRUYCK, N.: Inleiding to algemeene geographie. Amsterdam 1740.
9. REUSS, J.D.: Repertorium comentationum. Goetingae 1804, V. 5.
10. PINGRÉ, P.: Annales célestes (ed. BIGOURDAN). Paris 1901.
11. GINZEL, F.K.: Spezieller Kanon de Finsternisse. Berlin 1899.
12. CHAMBERS, G.F.: The story of eclipses. London 1899.
13. PRANTL, E.: Aristotelis de coelo libri, IV (1877) 21, Lipsiae.
14. HEATH, T.: Aristarchos of Samos. London 1913.
15. WOLF, R.: Geschichte der Astronomie, S. 154. München 1877.
16. WOLF, R.: Geschichte der Astronomie, S. 48. München 1877.
17. LANGRENUS, M.F.: Tractatus de vera longitudine. Antuerpiae 1644.
18. DOUBLET, E.: Histoire de l'astronomie, p. 279. Paris 1922.
19. LICETUS, T.: Litheophosphorus, p. 247. Utini 1640.
20. FRISCH, CH.: Joh. Kepleri opera omnia, 2, p. 297. Frankfurt 1858.
21. DU SÉJOUR, D.: Traité des mouvements apparents, I, p. 665. Paris 1786.

1.2. Photometrical Theory of the Umbra

1.2.1. Eclipse Scene

The general eclipse scene is given in Fig. 1.2.1 where S is the Sun, E the Earth and M the Moon. For simplication we have used the planes I and II in place of the Sun and the Moon. It is, therefore, important to see which is the error in angles r and γ involved by this approximation.

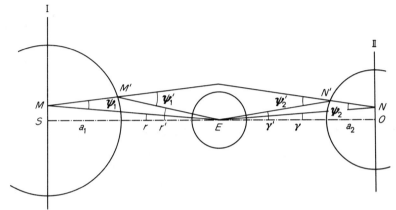

Fig. 1.2.1. General eclipse scene

The error in the angle r is given by

$$r' - r = \psi'_1 - \psi_1.$$

We have further

$$\operatorname{tg} \psi_1 = \frac{a + h'_0}{SE}$$

and by differentiation

$$\frac{d\psi_1}{\operatorname{tg} \psi_1} = \frac{MM'}{SE} \cos^2 \psi_1.$$

As the greatest value of MM' can be $MM' \leq a_1$ we get ($\cos^2 \psi \approx 1$)

$$\frac{d\psi_1}{\operatorname{tg} \psi_1} = \frac{d\psi_1}{\psi_1} \leq \frac{a_1}{SE} = R_\odot = 4\tfrac{1}{2}\%.$$

This error is negligible.

In the same way we get for $\gamma' - \gamma$

$$\frac{d\psi_2}{\psi_2} \leq \frac{a_3}{EM} = R_{\mathrm{C}} = 4\tfrac{1}{2}\%.$$

This error may be of minor importance in detailed research, but fortunately enough our measurements are frequently limited to the limb of the Moon, where $N \equiv N'$.

Consequently we are concerned only with solar (I) and lunar (II) planes and for the angles we have

$$\sin \psi_1 = \frac{a+h_0}{SM}, \quad \psi_1 = \pi_\odot \left(1 + \frac{h'_0}{a}\right)$$
$$\sin \psi_2 = \frac{a+h_0}{EM}, \quad \psi_2 = \pi_\mathrm{C} \left(1 + \frac{h'_0}{a}\right) \quad (1.10)$$

and finally

$$r = (\pi_\mathrm{C} + \pi_\odot)\left(1 + \frac{h'_0}{a}\right) - \omega \quad (1.11)$$

where ω is the deflection of the ray by the refraction.

1.2.2. General Photometrical Theory of the Umbra

The simplified eclipse scene is given in the Fig. 1.2.2, where I is the solar or II the lunar plane and E the center of the Earth. The solar illumination at the given point N of the lunar plane is produced by rays

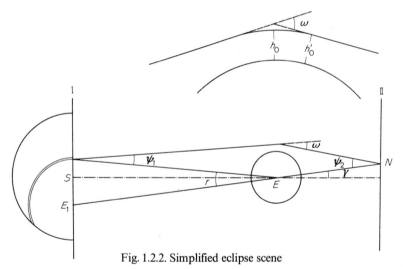

Fig. 1.2.2. Simplified eclipse scene

forming a kind of cone whose base is the solar disk and the point N its vertex. In this cone the individual solar rays are deviated and attenuated at different rates depending only on their minimum height h_0. The luminous flux transported by an elementary pencil of rays depends also

Photometrical Theory of the Umbra

on the place on the solar disk whence it issues. This means simply that the computation of the illumination e must respect both these variable factors in the whole cone [1].

The mathematical method used to do this is integration over the whole solar disk of the form

$$e = \int T(h_0)\, dq \qquad (1.12)$$

where $T(h_0)$ is the attenuation or transmission coefficient of the solar rays passing at the minimal or grazing height h_0 and dq the elementary luminous flux issued from one point of the solar disk.

Outside the eclipse we have $T(h_0) = 1$ and hence (1.2.3)

$$E = \int dq = \pi R_\odot^2 \left(1 - \frac{\kappa}{3}\right). \qquad (1.13)$$

The density of the shadow will be given by

$$D = \log_{10} E - \log_{10} e \qquad (1.14)$$

and the numerical connection between e/E and D may be seen in the following Table 1.5.

Table 1.5

Density D	0.0	0.5	1.0	1.5	2.0	2.5	3.0
Intensity e/E	1.0	0.312	0.1	0.0312	0.01	0.00312	0.001

The comparison of both lines clearly shows the advantage of the density D over the ratio of illuminations e/E.

1.2.3. Integration in the Solar Plane

The simplest method of integration is to develop the above given form of the integral [Eq. (1.12)] in order to proceed in the solar plane. As the transmission coefficient $T(h_0)$ depends on h_0, we chose the first integration with the constant value of h_0. The integration element is, therefore, an elementary ring circumscribed from the point E_1 (Fig. 1.2.3) by the radius r so that its brightness will be

$$di = 2 \int_0^{\varepsilon_0} b(R)\, r\, dr\, d\varepsilon \qquad (1.15)$$

where $b(R)$ is the brightness of the spot on solar disk at the distance R from its center. Generally we may assume for it the formula

$$b(R) = 1 - \kappa + \frac{\kappa}{R_\odot} \sqrt{R_\odot^2 - R^2} \qquad (1.16)$$

and the integration [Eq. (1.15)] can be carried out analytically. The above relation given for $b(R)$ holds good up to about 97% R and closer to the limb the observed brightness is less than its computed value, as was shown, for instance, by HEYDEN [2] during the solar eclipse.

There are two reasons which make the determination of $b(R)$ difficult at the extreme limb, namely the scintillation and diffused parasitic light in the atmosphere and in the instrument. Therefore, JULIUS [3] proposed the use of photometrical measurements of nearly total eclipses of the

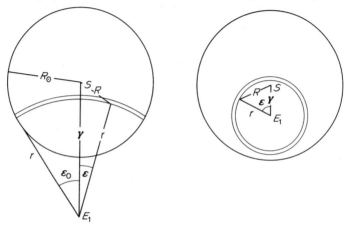

Fig. 1.2.3. Integration in the solar plane

Sun where the lunar disk screens progressively different parts of the Sun without any harmful interference from instrumental or atmospheric perturbations. The Julius method was already used with success in different eclipses.

The integration of di depends upon the position of the point E_1 with regard to the solar disk.

a) For $\gamma \geq R_\odot$ (Fig. 1.2.3 left) we have

$$di = 2\int_0^{\varepsilon_0} (1-\kappa)\, dr\, d\varepsilon + 2\int_0^{\varepsilon_0} \frac{\kappa r}{R_\odot} \sqrt{2r\gamma} \sqrt{\cos\varepsilon - \cos\varepsilon_0}\, dr\, d\varepsilon$$

or

$$di = 2\left[(1-\kappa)\varepsilon_0 r + \frac{\kappa r}{R_\odot}\sqrt{2r\gamma}\, I_m\right] r\, dr \qquad (1.17)$$

with

$$I_m = \int_0^{\varepsilon_0} \sqrt{\cos\varepsilon - \cos\varepsilon_0} = \frac{\pi m \sqrt{2}}{8}\left(1 + \frac{m}{32} + \frac{3m^2}{1024} + \cdots\right), \qquad (1.18)$$

$$m = 4\sin^2\frac{\varepsilon_0}{2} = \frac{(R_\odot + \gamma - r)(R_\odot + r - \gamma)}{r\gamma}. \qquad (1.19)$$

b) For $\gamma < R_\odot$ and $r < R_\odot - \gamma$ we have (Fig. 1.2.3 right) we have first

$$R^2 = \gamma^2 + r^2 - 2r\gamma \cos \varepsilon,$$

$$b(R) = 1 - \kappa + \frac{\kappa}{R_\odot^2} \sqrt{R_\odot^2 - (\gamma-r)^2 - 4r\gamma \sin^2 \frac{\varepsilon}{2}}$$

and consequently

$$di = 2\pi r(1-r) dr \qquad (1.20)$$

$$+ \frac{\kappa}{R_\odot} \sqrt{R_\odot^2 - (\gamma-r)^2} \int_0^\pi \sqrt{1 - \frac{4r\gamma}{R_\odot^2 - (\gamma-r)^2} \sin^2 \frac{\varepsilon}{2}} \, d\varepsilon \, dr$$

where the integral is an elliptic integral of the second kind

$$\int_0^\pi \sqrt{1 - \frac{4r\gamma}{R_\odot^2 - (\gamma-r)^2} \sin^2 \frac{\varepsilon}{2}} \, d\varepsilon = 2E\left(\vartheta, \frac{\pi}{2}\right) \qquad (1.21)$$

with

$$\sin^2 \vartheta = \frac{4r\gamma}{R^2 - (\gamma-r)^2} \qquad (1.22)$$

to be found in tables [4].

Both cases a) and b) may be written by the same formula

$$di = (P - \kappa Q) dr \qquad (1.23)$$

with the following definitions

a)
$$P = \varepsilon_0 r$$
$$Q = P - \frac{r}{R_\odot} \sqrt{2r\gamma} \, I_m. \qquad (1.24)$$

b)
$$P = 2\pi r$$
$$Q = P - \frac{2}{R_\odot} \sqrt{R_\odot^2 - (\gamma-r)^2} \, E\left(\vartheta, \frac{\pi}{2}\right). \qquad (1.25)$$

The values of $P/2$ and $Q/2$ are given in the Table 1.6 [4].

For some investigations, in particular at the edge of the shadow, it is necessary to allow for the actual luminance at the extreme solar limb. For this purpose we may use the values of $b(R)$ determined by HEYDEN [2] in the interval $99\% < R < 99.9\%$ and join it to the classical values for $R < 97\%$ obtained by ABBOT [5] and satisfying the adopted

18 Lunar Eclipses

Table 1.6

$r-\gamma$	$\gamma=0'$		$\gamma=5'$		$\gamma=10'$		$\gamma=20'$		$\gamma=30'$	
(')	$P/2$	$Q/2$	$P/2$	$Q/2$	$P/2$	$Q/2$	$P/2$	$Q/2$	$P/2$	$Q/2$
+16	50.3	50.3	0.0	0.0	0.0	0.0	0.0	0.0	0.0	0.0
15	47.1	30.7	11.3	8.2	8.8	6.4	7.9	5.9	6.8	5.0
14	44.0	22.7	15.5	9.6	12.1	7.5	10.1	6.3	9.4	5.8
13	40.8	17.0	18.5	10.1	14.4	7.8	12.0	6.5	11.2	6.1
12	37.7	12.8	20.8	12.2	16.0	7.7	13.5	6.5	12.6	6.0
11	34.6	9.5	22.6	10.0	17.3	7.5	14.6	6.3	13.7	5.9
10	31.4	6.9	24.2	9.9	18.3	7.2	15.5	6.0	14.5	5.6
9	28.3	4.9	25.5	9.7	19.0	6.8	16.1	5.7	15.2	5.3
8	25.1	3.4	26.9	9.9	19.5	6.4	16.6	5.3	15.7	5.0
7	22.0	2.2	28.6	10.6	19.8	6.0	17.0	4.9	16.1	4.7
6	18.8	1.4	34.6	14.2	20.0	5.6	17.2	4.8	16.4	4.3
5	15.7	0.8	31.4	9.9	20.1	5.4	17.3	4.5	16.6	4.2
4	12.6	0.4	28.3	7.2	20.0	5.2	17.3	4.2	16.7	4.0
3	9.4	0.2	25.1	5.1	19.8	4.9	17.3	4.0	16.6	3.8
2	6.3	0.0	22.0	3.6	19.5	4.8	17.0	3.8	16.6	3.7
+ 1	3.1	0.0	18.8	2.5	19.0	4.6	16.8	3.7	16.4	3.6
0	0.0	0.0	15.7	1,7	18.5	4.6	16.5	3.6	16.2	3.5
− 1			12.6	1.1	18.0	4.6	15.8	3.6	15.9	3.5
2			9.4	0.7	17.5	5.0	15.5	3.5	15.5	3.5
3			6.3	0.4	17.1	5.5	15.0	3.5	15.1	3.5
4			3.1	0.2	16.8	5.2	14.3	3.5	14.6	3.5
5			0.0	0.0	15.7	5.0	13.6	3.5	14.0	3.7
6					12.6	3.5	12.9	3.6	13.4	3.7
7					9.4	2.5	12.0	3.6	12.7	3.8
8					6.3	1.5	11.1	3.6	12.0	3.9
9					3.1	0.7	10.2	3.6	11.2	3.9
10					0.0	0.0	9.1	3.6	10.3	4.0
11							7.8	3.4	9.3	4.0
12							6.9	3.3	8.3	4.2
13							5.6	3.0	7.1	3.8
14							4.3	2.7	5.7	3.5
15							2.8	2.0	3.9	2.9
−16							0.0	0.0	0.0	0.0

formula [Eq.(1.16)]. By numerical integration we then get di. It can be shown in addition that at the edge of the shadow the values of di depend very little on γ being principally the function of $\gamma-r+R_\odot$.

For $\lambda=5400$ Å we have got in comparison with classical values computed by Eq. (1.17) following numbers [6]

$\gamma-r+R_\odot$		0'	1'	2'	3'	4'
di	Eq. (1.17)	0.0	3.6	5.5	7.1	8.4
	Heyden [6]	0.0	3.3	5.4	7.0	8.4

Table 1.6

γ=36'		γ=40'		γ=45'		γ=50'		γ=58'		γ=68'	
P/2	Q/2	P/2	Q/2	P/2	Q/2	P/2	Q/2	P/2	Q/2	P/2	Q/2
0.0	0.0	0.0	0.0	0.0	0.0	0.0	0.0	0.0	0.0	0.0	0.0
6.6	4.8	6.5	4.7	6.4	4.6	6.3	4.6	6.2	4.5	6.1	4.5
9.1	5.7	9.0	5.6	8.9	5.5	8.8	5.4	8.6	5.3	8.5	5.3
10.9	5.9	10.8	6.0	10.6	5.7	10.5	5.6	10.3	5.5	10.2	5.4
12.2	5.9	12.1	5.8	11.9	5.7	11.8	5.7	11.6	5.6	11.5	5.5
13.3	5.7	13.2	5.7	13.0	5.6	12.8	5.4	12.7	5.3	12.5	5.3
14.2	5.5	14.0	5.4	13.8	5.3	13.7	5.3	13.5	5.2	13.4	5.2
14.9	5.2	14.7	5.2	14.5	5.1	14.4	5.0	14.2	5.0	14.1	4.9
15.4	4.9	15.2	4.9	15.1	4.8	14.9	4.8	14.8	4.7	14.6	4.7
15.8	4.7	15.7	4.6	15.5	4.6	15.4	4.5	15.2	4.5	15.1	4.4
16.1	4.4	16.0	4.4	15.8	4.3	15.7	4.3	15.6	4.2	15.4	4.2
16.3	4.2	16.2	4.1	16.1	4.1	15.9	4.0	15.8	4.0	15.7	4.0
16.5	4.0	16.3	3.9	16.2	3.9	16.1	3.8	16.0	3.8	15.9	3.8
16.5	3.8	16.4	3.8	16.3	3.8	16.2	3.7	16.1	3.7	16.0	3.7
16.4	3.6	16.4	3.6	16.3	3.6	16.2	3.6	16.1	3.6	16.1	3.6
16.3	3.6	16.3	3.5	16.2	3.5	16.1	3.5	16.1	3.5	16.1	3.5
16.1	3.5	16.1	3.5	16.1	3.5	16.0	3.4	16.0	3.4	16.0	3.4
15.9	3.5	15.9	3.4	15.9	3.5	15.8	3.4	15.8	3.4	15.9	3.4
15.6	3.5	15.6	3.5	15.6	3.5	15.6	3.4	15.6	3.4	15.6	3.5
15.2	3.5	15.2	3.5	15.3	3.5	15.2	3.5	15.3	3.5	15.4	3.5
14.7	3.6	14.8	3.6	14.9	3.6	14.8	3.6	14.9	3.6	15.0	3.6
14.2	3.6	14.3	3.7	14.4	3.7	14.4	3.7	14.5	3.7	14.6	3.7
13.7	3.7	13.8	3.8	13.9	3.8	13.9	3.8	14.0	3.8	14.2	3.9
13.0	3.8	13.1	3.9	13.3	3.9	13.3	3.9	13.5	4.0	13.6	4.0
12.3	3.9	12.5	4.0	12.6	4.0	12.7	4.1	12.9	4.1	13.0	4.2
11.5	4.0	11.7	4.1	11.9	4.2	12.0	4.2	12.1	4.4	12.3	4.3
10.7	4.2	10.9	4.2	11.0	4.3	11.2	4.3	11.4	4.4	11.6	4.5
9.7	4.2	9.9	4.3	10.1	4.4	10.3	4.3	10.5	4.4	10.7	4.5
8.7	4.2	8.9	4.3	9.1	4.3	9.2	4.4	9.4	4.5	9.6	4.6
7.5	4.1	7.7	4.2	7.9	4.3	8.0	4.3	8.2	4.4	8.4	4.5
6.0	3.7	6.3	3.9	6.4	4.0	6.6	4.1	6.7	4.2	6.9	4.3
4.3	3.1	4.4	3.2	4.5	3.3	4.6	3.4	4.8	3.5	4.9	3.6
0.0	0.0	0.0	0.0	0.0	0.0	0.0	0.0	0.0	0.0	0.0	0.0

1.2.4. General Transmission Coefficient

Terrestrial atmosphere produces a multiple effect on the intensity of the light traversing it. We may distinguish the following components:

a) The *molecular scattering* of light should be considered first of all. Its theory has been given by RAYLEIGH [7] and its validity is limited to the pure atmosphere known also under the name of the *Rayleigh atmosphere*.

b) The *scattering on aerosols* always present in the actual atmosphere. Its theory has been developed by MIE and others [8]. Both molecular and aerosol scattering have an attenuation effect on the light by removing a part of the light flux from the primitive direction and throwing it in all directions.

c) The *true absorption*, producing in the spectrum the discrete absorption bands. A part of light flux is transformed in some other kind of energy e.g. in thermal energy of the absorbing molecules.

The combined effect of these three factors is named commonly as the *extinction* of light.

d) The *refraction* changes the primitive structure of the light beam by modifying its normal divergence and attenuating, therefore, the illumination of the lunar plane.

The complex action of these four factors contributes to the general transmission coefficient $T(h_0)$ which depends on the atmospheric structure.

1.2.5. Molecular Scattering of Light

The apparent absorption coefficient of the pure air is expressed by RAYLEIGH's law and completed by a term of CABANNES [7]

$$A = \frac{32\pi^3 \log e (\mu-1)^2}{3\lambda^4 N} \frac{6+3\delta}{6-7\delta} \tag{1.26}$$

where N is the numbre of molecules in 1 cm³,
λ the wave-lenght of the light
μ its index of refraction
δ the depolarisation factor of CABANNES.

For 1 km of the air having the density $\rho = 1$ (NTP) we obtain

$$A_1 = 5.71 \times 10^{-13} (\mu-1)^2 / \lambda^4 \tag{1.27}$$

or for the height of the homogenious atmosphere ($=8$ km)

$$A_8 = 4.57 \times 10^{-12} (\mu-1)^2 / \lambda^4. \tag{1.28}$$

As the consequence of λ^{-4} in the above formulae, the molecular scattering of light and the absorption coefficients are highly selective. From the violet (4000 Å) to the red (7000 Å) end of the spectrum the values of A_1 and A_8 change in the ratio of about 10:1 as it is shown in the Table 1.7.

The great selectivity of the molecular scattering is the reason for the red color characteristic of lunar eclipses.

The attenuation of light by the molecular scattering is

$$10^{-AM(h_0)} \tag{1.29}$$

Table 1.7

λ (Å)	A_8	λ (Å)	A_8
4000	0.1578	6000	0.0300
4200	0.1288	6200	0.0263
4400	0.1064	6400	0.0232
4600	0.0889	6600	0.0204
4800	0.0745	6800	0.0181
5000	0.0630	7000	0.0161
5200	0.0537	7500	0.0122
5400	0.0460	8000	0.0094
5600	0.0397	8500	0.0074
5800	0.0345	9000	0.0059

where $M(h_0)$ is the air mass traversed by the rays during their travel in the atmosphere. The computation of this very important quantity will be given later (1.2.13).

1.2.6. Aerosol Scattering

The actual atmosphere is polluted by many kinds of aerosols of different origins. In low layers (troposphere) these aerosols are mainly of terrestrial origin and of inorganic or organic nature. Their concentration decreases rapidly with height and the tropopause is for various reasons considered the natural limit or barrier for the main bulk of the terrestrial aerosols. The pollution of the upper atmosphere is, however, mainly of meteoric origin.

The scattering of light on the aerosols, a difficult problem, has received the attention of many authors beginning with MIE and a very good monograph on these problems has been written by VAN DE HULST [8]. The general conclusion of these studies is that the spectral selectivity of the aerosols is less than the molecular scattering.

In view of their varied nature and changing distribution, it is very difficult to take into account the influence of the terrestrial aerosols. Some information can be obtained from the direct determination of atmospheric transmission by subtracting from it the theoretical value given by the formula [Eq. (1.27)] of RAYLEIGH-CABANNES. The difference — if positive— can be attributed to the aerosols in these spectral regions where the true absorption can be neglected.

1.2.7. Astronomical Determinations

Astronomical determinations of the atmospheric transmission are based on Bouguer's graph. By some exact photometric method we measure the monochromatic intensity of any extraterrestrial source like the Sun or the stars in different zenithal distances z. If now $M(z)$ is the air mass traversed by the rays in the Rayleigh atmosphere and $H(z)$ the corresponding quantity for aerosols, we have the formula (Fig. 1.2.4)

$$\log I = \log I_0 - AM(z) - CH(z) \tag{1.30}$$

where I_0 is the extraterrestrial intensity of the source and A or C the absorption coefficients of the air and of aerosols. If we put at zenith

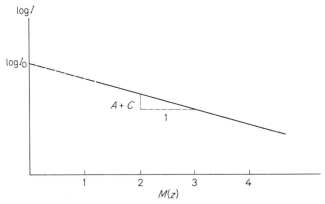

Fig. 1.2.4. Bouguer's straight line

$M(z) = H(z) = 1$, we have for $z < 75°$ with good approximation $M(z) = H(z) = \sec z$ and our formula will be now

$$\log I = \log I_0 - (A + C) \sec z \tag{1.31}$$

that of Bouguer's straight line whose slope gives the sum $A + C$.

Very extensive determinations of $A + C$ were performed some time ago during the measurements of the solar constant by ABBOT and his collaborators [9]. Their discussion will be given later (1.3.7).

1.2.8. Terrestrial Method

The terrestrial method is more difficult to carry out because of the small value of A. The simplest method consists of measuring the light intensities i_1 and i_2 of a point light source at two very different distances r_1 and r_2. We have

$$\frac{i_1}{i_2} = \left(\frac{r_2}{r_1}\right)^2 \cdot 10^{A(r_2 - r_1)} \tag{1.32}$$

which leads to the value of A

$$A = \frac{\log i_1 - \log i_2 - 2(\log r_2 - \log r_1)}{r_2 - r_1}. \tag{1.33}$$

To reduce the influence of measurement errors the difference $r_2 - r_1$ should be as great as is compatible with the great value of the ratio i_1/i_2 involved by this choice.

In these measurements [10, 11] a powerful electric bulb (Philips 1 or 5 kw) was used. Its luminous intensity was controlled by means of a variable resistance and the ammeter maintaining the constant current intensity. In the first series a visual stellar photometer with the artificial comparison star gave better results than a plage photometer because of greater sensibility to the weak light fluxes and relatively slight sensibility to the scintillation. The spectral selection was obtained by means of Wratten gelatine filters.

Two other series were photographic using the principle of the objective prism. The spectrum of the remote source on the film was accompanied by several comparison spectra of graduated intensity produced by a constant light source incorporated into the photometer box. The densitometric measurements of both kinds of spectra on the film gave the necessary data for the Eq. (1.38).

Both visual and photographic methods were supplemented by the density determination of the neutral filter that had to be used at the near observing station (r_1) in order to reduce the great light flux and approximate it to the value at the distant station (r_2). This determination was performed in the laboratory on the optical bench.

The values of A, of course, depend on the atmospheric conditions i.e. on the atmospheric haze present almost at all times in the troposphere. They are, therefore, greater than the theoretical valeurs computed in the Rayleigh atmosphere [Eq. (1.27)]. Some of these results are given in the following Table 1.8.

Table 1.8

	2830 m [10]		2543 m [11]		624 m [11]	
Mean height						
Mean air density	0.70		0.74		0.93	
Length of the basis	30 km		62 km		35 km	
Å	a_t	a_{min}	a_t	a_{min}	a_t	a_{min}
4600	0.0078	0.0084				
5000			0.0058	0.0080		
5300	0.0044	0.0052	0.0046	0.0076		
5500			0.0040	0.0070	0.0050	0.0222
5900			0.0030	0.0060	0.0038	0.0179
6300	0.0022	0.0036	0.0023	0.0051	0.0029	0.0162

a_t theoretical value in the Rayleigh atmosphere/km; a_{min} minimal measured value/km.

1.2.9. Discussion of A

We have at our disposal three methods that give the value of the extinction coefficient A:

i) *Theoretical* value according to the Rayleigh-Cabannes formula which is strictly valid only for a pure atmosphere.

ii) *Astronomical* value obtained by the Bouguer's graph where the pollution is partially included.

iii) *Terrestrial* value obtained on the horizontal basis-line where the atmospheric pollution intervenes in full weight.

Our choice depends, of course, on the aim that we ascribe to the photometrical theory.

If the theory should be the truest description of the natural phenomenon of the eclipse, the terrestrial value (iii) is best fitted to this purpose. However, we need the values of A determined on horizontal basis lines of increasing heights h_0. The realization of such an undertaking seems to be very difficult above $h_0 = 4-5$ km.

If, on the other hand, the theory should be a tool for the exploration of unknown properties of the atmosphere, the theoretical value of A is to be preferred. From the differences between the theory and observations we may then reach a conclusion on the new feature of the atmospheric structure, especially in its upper parts.

Different photometrical theories have made the choice according to the scientific level at the time. HEPPERGER and SEELIGER have adopted the astronomical value of A known only at the end of the 19th century. FESENKOV adopted a compromise between the astronomical and terrestrial method adopting for A the value obtained by astronomical observations by MÜLLER at Säntis (2500 m) near the horizon, i.e. where the hazy layer is greatly involved. SAUSSURE takes at first the theoretical value corrected by Fowle's method in respect to the water vapor content in different layers determined by LANGLEY. Finally, we have deliberately chosen the first method, adopting the theoretical values of A without any correction.

1.2.10. Attenuation of Light by Refraction

Refraction gives rise to the attenuation of solar rays by its differential character. An elementary pencil of rays (Fig. 1.2.5) issuing from the solar element at M and contained between the angles τ and $\tau + d\tau$, illuminates an elementary ring of the area of dS' in the lunar plane. If there were no refraction, the illuminated ring would be at dS. Since in both cases the same light flux illuminates the lunar plane, the attenuation of illumination will be

$$\frac{dS}{dS'} = \Phi^{-1}. \tag{1.34}$$

From the geometry of the scene we obtain, with regard to the smallness of the angles [1]

$$\Phi = \frac{dS'}{dS} = \left[1 - \frac{\omega}{\pi_{\mathbb{C}} + \pi_{\odot}}\left(1 - \frac{h'_0}{a}\right)\right]\left[1 - a\frac{d\omega}{dh'_0}\frac{1}{\pi_{\mathbb{C}} + \pi_{\odot}}\right]$$
$$= \frac{r}{\psi}\left(1 - a\frac{d\omega}{dh'_0}\frac{1}{\pi_{\mathbb{C}} + \pi_{\odot}}\right). \tag{1.35}$$

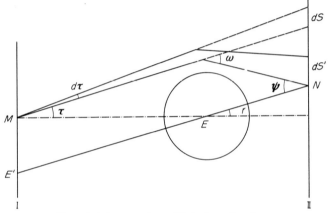

Fig. 1.2.5. Attenuation of light by refraction

It should be noted that this formula is symmetrical in the distances Sun–Earth and Earth–Moon or in the parallaxes $\pi_{\odot}$ and $\pi_{\mathbb{C}}$. The extinction (1.29) has also the same property. In other words the light element on the Sun and the illuminated element on the Moon can be exchanged together without the general transmission coefficient changes.

Further, we may note that the expression of Φ consists of two terms

$$\frac{r}{\psi} \tag{1.35a}$$

and

$$1 - a\frac{d\omega}{dh'_0}\frac{1}{\pi_{\mathbb{C}} + \pi_{\odot}} \tag{1.35b}$$

which have an opposite action. This can be understood in the following geometrical way. Let us consider an elementary cylindrical beam traversing the atmosphere (Fig. 1.2.6). With uniform refraction in height, the ring ds in the lunar plane will be illuminated. Due to its smaller radius, its area will be smaller than the area dS and, therefore, an increase of illumination would result according to the first term given above. The second term expresses the enlargement of the ring to dS' by differential refraction and consequently the attenuation of illumination.

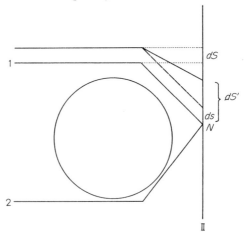

Fig. 1.2.6. Two components of the attenuation by refraction

If we follow the variation of Φ with decreasing altitude h, these is in the upper atmosphere an appreciable attenuation in spite of the small value of the refraction ω (Table 1.9). In the troposphere Φ reaches its maximum, then decreases to the value of $\Phi = 1$ at the height where the condition

$$r = \frac{\psi}{1 - a\dfrac{d\omega}{dh'_0} \dfrac{1}{\pi_\odot + \pi_\mathrm{C}}} \qquad (1.35\,\mathrm{c})$$

is satisfied. That takes place approximately at $h = 2$ km. The lower rays are, on the contrary, amplified and for $r = 0$ there is a focussing effect. The corresponding height of the ray is given by the condition

$$\omega = (\pi_\mathrm{C} + \pi_\odot)\left(1 + \frac{h_0}{a}\right). \qquad (1.35\,\mathrm{d})$$

Lower down the value of Φ becomes negative. That has only a geometrical meaning but indicates also that the point N of the lunar plane receives double illumination (Fig. 1.2.7). The first illuminating beam traverses the atmosphere above the altitude where $\Phi = 1$ and the second beam 2 below this altitude and at the antipodal point of the

Earth. However, the action of the second beam, as it passes very low where the extinction is great and the cloudiness frequent, is almost negligible.

Finally, we should emphasize that the attenuation by refraction is nearly neutral and the shadow due only to this would be approximately grey.

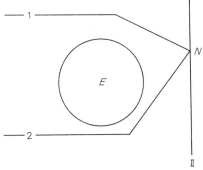

Fig. 1.2.7. Double illumination in the shadow

1.2.11. Auxiliary Shadow

Having computed the general transmission coefficient in Rayleigh atmosphere i.e.

$$T(h_0) = 10^{-AM(h_0)} \Phi^{-1} \qquad (1.36\,\text{a})$$

or in form of optical density

$$\log T(h_0) = D'(h_0) = -AM(h_0) - \log \Phi \qquad (1.36\,\text{b})$$

we have at the same time the density of the *auxiliary shadow* thrown by a solar element on the lunar plane. As a numerical example of this kind of calculation, it is given in the following table for two colors blue (4600 Å) and red (6200 Å) and for the lunar parallax $\pi_{\mathbb{C}} = 57'$.

By the examination of this table we conclude that in the outer parts of the shadow only the refraction term Φ^{-1} is of importance, giving there the neutral color of these regions. On the other hand, the internal part of the shadow is greatly influenced by the extinction term AM which is very selective. The inner part of the shadow is, therefore, red.

In general, we assume for the calculation of the auxiliary shadow the mean yearly structure of the atmosphere at middle latitudes and we take into account only the variations of the lunar parallax. However, the table of the auxiliary shadow can be computed for any climatic and geographic situation on the basis of the Dioptric Tables of the Earth's atmosphere [12].

Table 1.9

h_0 (km)	B—4600 Å				R—6200 Å			
	$r(')$	d_1	d_2	D'	$r(')$	d_1	d_2	D'
2	5.74	5.21	1.39	6.90	6.35	1.54	1.72	3.26
3	9.27	4.65	1.94	6.59	9.84	1.38	1.95	3.33
4	14.03	4.14	2.15	6.29	14.54	1.22	2.16	3.38
5	18.18	3.66	2.20	5.89	18.64	1.08	2.24	3.32
6	22.53	3.25	2.28	5.53	22.94	0.96	2.28	3.24
7	25.86	2.88	2.23	5.11	26.23	0.85	2.23	3.08
8	28.74	2.54	2.20	4.74	29.07	0.75	2.20	2.95
9	31.13	2.24	2.15	4.39	31.44	0.66	2.15	2.81
10	33.22	1.96	2.22	4.18	33.50	0.58	2.22	2.80
12	38.05	1.48	2.13	3.61	38.28	0.44	2.12	2.56
14	44.30	1.09	2.26	3.35	41.49	0.32	2.26	2.58
16	44.80	0.78	2.19	2.97	44.95	0.23	2.19	2.42
18	48.87	0.55	2.20	2.75	48.97	0.16	2.20	2.36
20	51.39	0.39	2.02	2.41	51.46	0.12	2.01	2.13
22	53.13	0.29	1.89	2.18	53.18	0.08	1.89	1.97
24	54.33	0.21	1.71	1.92	54.37	0.06	1.71	1.77
26	55.15	0.15	1.58	1.73	55.17	0.05	1.57	1.62
28	55.74	0.11	1.45	1.56	55.76	0.03	1.45	1.48
30	56.19	0.08	1.31	1.39	56.20	0.02	1.30	1.32
35	56.87	0.04	1.03	1.07	56.88	0.01	1.03	1.04
40	57.21	0.02	0.69	0.71	57.21	0.01	0.69	0.70

$d_1 = -\log \Phi$, $d_2 = -AM(h_0)$, $D' = d_1 + d_2$.

The density of the actual shadow is computed according to the formula [Eq. (1.12)] by integration over the interval of $2R_\odot$ of the auxiliary shadow. The above discussed differences of color are also conserved in the actual shadow, even when there is less contrast, and the observations confirm it clearly.

As the value of the transmission coefficient $T(h_0)$ is independent of the direction of the light rays [Eq. (1.36)], this integration can be formulated as giving the illumination proportional to the illumination at the subsolar point by the diffusing solar disk, itself illuminated by a point light source placed on the spot of the lunar plane where we compute the density of the actual shadow. The obscuration of the solar disk can be accounted for by adopting the variable albedo of the diffusing solar disk. This formulation of our problem was first given by GLEICHEN [13].

1.2.12. The Normal Densities of the Shadow

The normal densities of the shadow were computed some years ago [1] for the Rayleigh atmosphere at middle latitudes in following spectral regions:

Photometrical Theory of the Umbra

Table 1.10

Color	Wave-length (Å)	$c=\mu-1$	A [Eq. (1.28)]	κ [Eq. (1.16)]
Red	6200	292.0×10^{-6}	0.0263	0.54
Green	5400	293.2×10^{-6}	0.0460	0.62
Blue	4600	295.5×10^{-6}	0.0889	0.73

and for three lunar parallaxes: 54′, 57′ and 61′.

In addition to the densities, we may calculate the so-called *mean air mass* M_0 which enables us to achieve a detailed comparison of the theory with observations. As the actual extinction coefficient A in the extinction term AM may be somewhat different from the theoretical value [Eq. (1.26)] we differentiate the expression for the illumination

$$e = \int_{h_1}^{h_2} 10^{-AM} \Phi^{-1} \, di \, dr \qquad (1.37)$$

giving us

$$\frac{de}{dA} = -2.30 \cdots \int 10^{-AM} \Phi^{-1} M \, di \, dr$$

or for the optical density

$$\frac{dD}{dA} = M_0 = \frac{\int 10^{-AM} \Phi^{-1} M \, di \, dr}{\int 10^{-AM} \Phi^{-1} \, di \, dr}. \qquad (1.38)$$

This expression having the dimension of the air-mass is, therefore, called the *mean air mass* and it is given together with the shadow density in the following Table 1.11.

For a deeper discussion it is useful to visualize the form of the integral of illumination [Eq. (1.37)]. This is done in the Fig. 1.2.8. The influence of different atmospheric layers and the extension of the effective shadow terminator is visible for different points of the shadow. The continuation of the curves to the negative values of the angle r corresponds to the double illumination of the inner shadow, as already (1.2.10) explained. The main use of these curves is to be found in the discussion of tropospheric perturbations of the shadow (1.3.9).

Let us discuss finally the influence of the parallax on the density of the shadow. At the edge of the shadow, where the isophotes of the auxiliary shadow are only slightly incurvated, this influence is very small if we adopt, instead of angle γ, as an argument the angular distance from the edge, defined by the relation

$$\gamma'' = \pi_{\mathbb{C}} - R_\odot - \gamma. \qquad (1.39)$$

Table 1.11

	$\pi_{☽}$	D						M_o					
		$\gamma=0'$	$5'$	$10'$	$20'$	$30'$	$35'$	$\gamma=0'$	$5'$	$10'$	$20'$	$30'$	$35'$
R 0.62 μ	54'	3.10	3.09	3.06	2.97	2.76	2.58	49.5	48.4	43.5	27.8	17.5	11.8
	55'	3.14	3.13	3.10	3.01	2.79	2.62	50.3	49.5	44.4	28.7	18.9	13.1
	56'	3.18	3.16	3.14	3.05	2.81	2.66	51.2	50.6	45.3	29.8	20.1	14.3
	57'	3.21	3.20	3.17	3.09	2.83	2.70	52.2	51.7	46.3	31.1	21.2	15.5
	58'	3.25	3.24	3.21	3.12	2.85	2.73	53.4	52.6	47.4	32.3	22.1	16.5
	59'	3.29	3.27	3.24	3.16	2.87	2.76	54.5	53.4	48.7	33.7	22.8	17.5
	60'	3.32	3.31	3.27	3.18	2.90	2.78	55.5	54.2	50.0	35.2	23.5	18.4
	61'	3.36	3.35	3.31	3.20	2.92	2.80	56.7	54.9	51.8	36.8	24.1	19.3
G 0.54 μ	54'	4.10	4.04	3.91	3.49	3.09	2.79	44.9	43.3	37.8	24.7	15.3	9.9
	55'	4.17	4.11	3.97	3.53	3.13	2.86	45.8	44.0	38.6	25.5	16.2	11.0
	56'	4.24	4.18	4.03	3.58	3.17	2.91	46.9	44.9	39.6	26.5	17.1	12.1
	57'	4.30	4.24	4.09	3.63	3.21	2.97	48.0	46.1	40.8	27.7	18.1	13.2
	58'	4.36	4.31	4.15	3.68	3.25	3.02	49.2	47.6	42.2	29.0	19.0	14.2
	59'	4.43	4.37	4.22	3.74	3.29	3.07	50.5	49.2	43.8	30.2	19.9	15.2
	60'	4.49	4.43	4.28	3.80	3.33	3.12	52.0	50.8	45.5	31.5	20.8	16.1
	61'	4.55	4.49	4.35	3.87	3.37	3.16	53.3	52.7	47.3	32.9	21.7	17.0
B 0.46 μ	54'	5.86	5.69	5.30	4.46	3.68	3.18	42.0	37.6	32.1	20.9	12.5	7.7
	55'	5.98	5.77	5.40	4.55	3.76	3.28	43.3	39.2	33.0	21.7	13.2	8.5
	56'	6.10	5.87	5.50	4.63	3.85	3.38	44.8	40.8	34.1	22.7	14.1	9.5
	57'	6.23	5.98	5.60	4.72	3.93	3.48	46.3	42.4	35.2	23.7	15.1	10.5
	58'	6.36	6.11	5.72	4.82	4.01	3.57	47.8	43.8	36.6	24.8	16.2	11.4
	59'	6.49	6.25	5.84	4.93	4.09	3.66	49.3	45.4	38.1	25.9	17.4	12.4
	60'	6.63	6.40	5.97	5.05	4.17	3.75	50.8	47.0	39.5	27.1	18.8	13.3
	61'	6.76	6.56	6.11	5.16	4.24	3.84	52.3	48.4	41.1	28.3	20.3	14.1

The densities with this new argument are given in the Table 1.12 and are valid only for the peripherial parts of the shadow.

As we penetrate towards the center, the influence of the parallax becomes greater and greater and reaches its maximum in the center of the shadow, as has already been stated by SEELIGER [14]. In the perigee

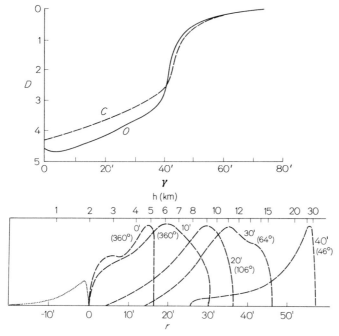

Fig. 1.2.8. Above: Density curves O observed and C computed of the shadow for green light and $\pi_{\mathbb{C}}=57'$. Below: Different forms of the illumination integrals for $\pi_{\mathbb{C}}=57'$. The figures at the curves give the distances γ and the angular extension of the terminator

the shadow is darker than in the apogee, as can be seen also geometrically. This influence of the parallax should be taken into account in every detailed discussion of the observed luminosities of lunar eclipses.

The mean air mass M_0 represents the influence of the Rayleigh atmosphere or the deviation of its extinction coefficient from the value adopted. In discussing other atmospheric factors, it is useful to consider the height of the maximum of the integration curves (Fig. 1.2.8) which represents some kind of effective ray. Any perturbation of transmission at this level has the greatest effect on the illumination in the lunar plane. In the outer part of the shadow, near its edge, the upper limit h_2 of the illumination integral [Eq. (1.37)] may serve the same purpose.

Table 1.12

	$\pi_{\mathbb{C}}$	$\gamma'' = 0'$	1'	2'	3'	4'	5'	6'	7'	8'	9'	10'
R 0.62 μ	54'	2.37	2.46	2.53	2.58	2.63	2.67	2.70	2.73	2.76	2.79	2.81
		7.3	9.2	10.7	11.8	13.0	14.0	15.2	16.3	17.5	18.5	19.5
	57'	2.35	2.46	2.52	2.58	2.63	2.67	2.70	2.73	2.75	2.78	2.81
		7.3	8.8	10.2	11.7	13.0	14.5	15.5	16.7	17.8	19.0	20.1
	61'	2.33	2.42	2.49	2.55	2.60	2.64	2.67	2.71	2.75	2.77	2.80
		7.3	8.3	9.5	10.6	11.7	13.1	14.5	15.7	17.0	18.4	19.3
G 0.54 μ	54'	2.50	2.62	2.72	2.79	2.86	2.93	2.99	3.04	3.08	3.13	3.17
		6.0	7.4	8.8	9.9	11.1	12.1	13.2	14.3	15.3	16.4	17.4
	57'	2.48	2.61	2.70	2.78	2.85	2.92	2.97	3.02	3.07	3.13	3.17
		5.7	7.1	8.4	9.7	11.0	12.1	13.2	14.3	15.3	16.3	17.2
	61'	2.46	2.58	2.68	2.76	2.83	2.90	2.95	3.01	3.07	3.12	3.16
		5.9	7.2	8.4	9.7	11.0	12.1	13.2	14.3	15.2	16.2	17.0
B 0.46 μ	54'	2.73	2.91	3.06	3.18	3.29	3.39	3.49	3.59	3.68	3.77	3.85
		4.1	5.4	6.6	7.7	8.6	9.5	10.5	11.4	12.5	13.4	14.4
	57'	2.71	2.89	3.04	3.16	3.28	3.38	3.48	3.58	3.67	3.76	3.84
		4.1	5.4	6.6	7.7	8.6	9.5	10.5	11.4	12.4	13.3	14.2
	61'	2.69	2.86	3.01	3.16	3.26	3.37	3.46	3.56	3.65	3.75	3.84
		4.3	5.4	6.4	7.5	8.5	9.5	10.5	11.4	12.4	13.3	14.1

The Table gives the densities D (above) and the air mass M_0 (below).

1.2.13. Theory of Refraction and Air Mass

From the preceding developments, it is quite clear that a knowledge of the refraction and of the air mass for every ray of the illuminating solar beam is of capital importance. The computation of these quantities belongs to classical astronomy, but in a different domain of arguments than in the eclipse theory.

In classical astronomy one is interested in the refraction and the air mass for oblique rays, i.e. for different zenithal distances z going away from the observer at a relatively low height. In the eclipse theory, however, we need these values for only $z = 90°$ but at different levels from the Earth's surface up to several tens of kilometers. Even when the basic formulae are the same in both cases, their further development is quite different. Therefore, the classical theory cannot be of any practical use for our purposes.

The fundamental equations for the refraction and the air mass are [12]

$$dR = -\frac{d\mu}{\mu}\,\text{tg}\,i, \tag{1.40}$$

$$dM = \rho \sec i\, dh \tag{1.41}$$

where the index of refraction $\mu = 1 + c\rho$ is a function of the air density ρ and the angle i between the ray and the radius vector is given by $(a=1)$

$$\sin i = 1 - (h - h_0) + c(\rho_0 - \rho). \tag{1.42}$$

The constant c ($\approx 3 \times 10^{-4}$) depends slightly on the wave-length.

For the integration of both formulae, we must know the aerological function $\rho = f(h)$ in some analytical or numerical form. In classical astronomy many authors have succeeded in approximating this function with some analytical form favourable to the integration. For our purposes none of these approximations is good enough and we are compelled, therefore, to use the numerical values of ρ and integrate the above formulae by a numerical method.

For this we introduce a new variable [1]

$$Z = \cos i \tag{1.43}$$

and we get

$$\omega = 2 \int_0^1 K\, dZ, \tag{1.44}$$

$$M = 2 \int_0^1 L\, dZ \tag{1.45}$$

3 Link, Eclipse Phenomena

where
$$K = c\rho\beta(1-c\rho\beta)^{-1}$$
$$L = K(c\beta)^{-1} \quad (1.46)$$
with
$$\beta = \left|\frac{d\rho}{dh}\right|\rho^{-1}.$$

The ratio $d\omega/dh_0$ may be obtained directly from the numerical value of ω. As actually the ratio $d\omega/dh'_0$ is needed, we multiply

$$\frac{d\omega}{dh'_0} = \frac{d\omega}{dh_0}\frac{dh_0}{dh'_0} = \frac{d\omega}{dh_0}(1-c\rho_0\beta_0)^{-1} \quad (1.47)$$

while the height of the asymptote h'_0 is given by

$$h'_0 = h_0 + c\rho_0. \quad (1.48)$$

The process leading to the table of numerical values of ρ and β as the function of the grazing height h_0 is as follows. We start from a table of numerical values of the air density ρ for each kilometer of height

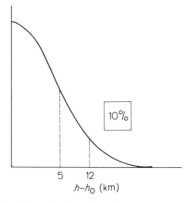

Fig. 1.2.9. Refraction integral

between 0 and 100 km. From these values we derive the gradient β, and we calculate the quantities K and L by means of Eq. (1.46). Then, for a series of minimal heights h_0 progressing by 1 km from 0 to about 40–50 km, we calculate the variable Z. The graphical representation of K and L as the function of Z gives the curves of integrals [Eq. (1.44)–(1.45)]. These curves (Fig. 1.2.9), rapidly decreasing with height, show that the atmospheric structure between h_0 and $h_0 + 12 - 15$ km has the greatest influence on the value of ω or M. In this interval already some 95% is achieved so that if the rest of the integrand is subject to an error of up to 20%, the error of the whole integral will be less than 1%. In other words the structure of the upper atmosphere above $h_0 + 15$ km enters in the integration only as a small correction term.

1.2.14. Confrontation of the Refraction Theory with Observations

It is not without interest to confront the refraction theory just outlined with direct determinations of the refraction, accompanied by direct measurements of the air density. We have carried out such work [15] at the Pic-du-Midi Observatory, its altitude enabling us to have at least three light trajectories of grazing heights $h_0 = 1.5$, 2.0 and 2.86 km, the last altitude being that of the Observatory itself.

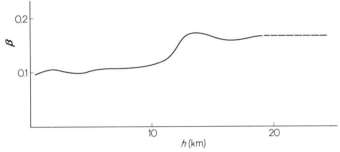

Fig. 1.2.10. Density gradient β at different altitudes

The direct determinations of the refraction were made on the rising or setting Sun by measuring its apparent zenithal distance with a theodolite and comparing it with the computed value according to the ephemeris and the moment of the observation.

The air density was computed according to the meteorological soundings at Nîmes and Bordeaux, going up to the altitude of 20–25 km which is sufficient to include about 95% of the refraction integral. Above this level we assumed the exponential extrapolation of the atmospheric structure. The air density was computed by

$$\rho = \frac{273}{1013} \frac{p}{T(1+0.61r)} \tag{1.49}$$

where p the pressure in millibars

T the temperature in °K

r the mixing ratio of the water vapor

at different altitudes.

To obtain the density gradient β we used the difference of $\log \rho$ between each kilometer plotted as the function of the height and graphically smoothed (Fig. 1.2.10).

The agreement between astronomical observations and the refraction computed [Eq. (1.44)] from aerological soundings is fairly good (Fig. 1.2.11) showing the correctness of the theory as well its applicability to the problem of the eclipses.

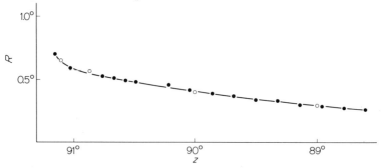

Fig. 1.2.11. Observed • and computed ○ refractions at the Pic-du-Midi Observatory

1.2.15. Climatic Influences on the Refraction and Air Mass

The course of the aerological function $\rho = f(h)$ depends upon the geographical latitude, the season of the year and the actual meteorological situation. Therefore, the values of ω and M will change with these factors. To compute it, we must start from the atmospheric structure determined for instance by the temperature distribution with altitude. Knowing this function $T(h)$, we have for the aerological function

$$\log \rho = \log \rho_0 + \log T_0 - \log T - \frac{434.3\, m}{k} \int_h^{h_g} \frac{dh_g}{T} \quad (1.50)$$

$$\beta = \frac{g}{T}\left(\frac{m}{k} + 10^{-3} \frac{dT}{dh_g}\right)$$

where T is the absolute temperature,
 ρ and the air density at the height h,
 T_0, ρ_0 the same quantities at the height h_0,
 m the molecular mass of the air,
 k the constant of gas,
 g the acceleration of gravity.

For the geopotential altitude we put

$$h_g = 10^{-3} \times g_0\, h - 1.54 \times 10^{-9}\, h^2. \quad (1.51)$$

This method was first used [16] with the meridional section of temperature isoplethes by RAMANATHAN [17] to compute ρ and β and from them the refraction and air mass for two extreme seasons of the year on both hemispheres. The results concerning the air density ρ are shown in Fig. 1.2.12 and those of the refraction and air mass in Fig. 1.2.13. To show

Photometrical Theory of the Umbra

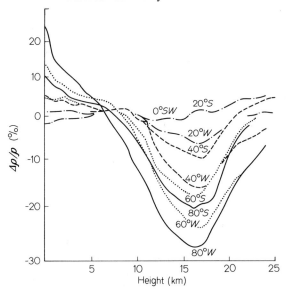

Fig. 1.2.12. Density variations with the height at different latitudes in winter (W) and summer (S)

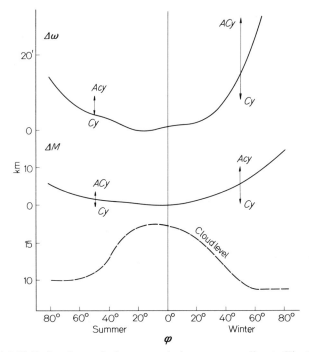

Fig. 1.2.13. Refraction and air mass variation corresponding to Fig. 1.2.12

the influence of the meteorological situation, the indications of the atmospheric structure during a cyclone and anticyclone given by RUNGE [16] were used for the same purpose. The influence of all these factors is not negligible.

The next similar attempt was carried out using the new International Geophysical Year data on meridional sections of the atmosphere up to 30 km for each month of 1958 between 70 S and 70 N. These charts [18] display the temperature distribution T as function of the pressure p. We have then between geopotential height h_g and pressure p the relation

$$h_g - h_{g_1} = 0.293 \int_{p_1}^{p} \frac{T}{p} dp \qquad (1.52)$$

which enables us to obtain by the numerical integration the dependence of the pressure upon the height. The corresponding air density is computed according to

$$\rho = \frac{273}{T} \frac{p}{1013}. \qquad (1.53)$$

We finally get the values of β by graphical treatment of log ρ. The values of ω and M were computed by the electronic computer ZUSE-Z-23 and constitute with other elements of the light path the Dioptric Tables of the Earth's atmosphere [12].

1.2.16. Climatic Variations of the Shadow Density

The normal densities of the shadow were calculated on the basis of the mean atmospheric structure at 45° latitude. Knowing the world distribution of the refraction and air mass, we are able to take into account these climatic influences on the shadow density.

As has already been stated the relation between the latitude φ on the shadow terminator and angle of position P on the lunar plane is as follows [Eq. (1.8)]

$$\cos P = \sin \varphi \sec \delta_\odot. \qquad (1.8)$$

By this formula it is also possible to determine for each date ($\delta_\odot$) and each latitude the corresponding angle of position. The series of latitudes progressing, for instance, by 10° around the Earth determines a fan of directions in the shadow. We compute for each of these latitudes the table of the auxiliary shadow, interpolate there for a series of densities 0.0; 0.1; 0.2 etc. the corresponding distances r from the center of the shadow and we construct finally the isophotes of the auxiliary shadow which are, of course, not circular, but deformated especially in the internal parts of the shadow.

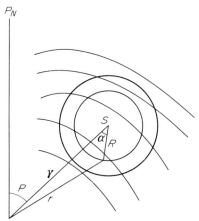

Fig. 1.2.14. General form of the integration in lunar plane

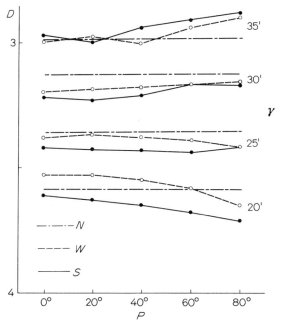

Fig. 1.2.15. Density variation in the shadow along the meridian in winter (W), summer (S), and in normal shadow (N) [19]

To compute the density of the actual shadow at any point of the lunar plane (Fig. 1.2.14) given by its polar coordinates P, γ, we place there the center of the fictive solar disk and we proceed to the integration

$$e = k \int_{\alpha=0}^{\alpha=2\pi} \int_{R=0}^{R=R_\odot} b(R)\, T(r)\, R\, d\alpha\, dR \qquad (1.54)$$

over the whole disk. The first integration may be accomplished (Fig. 1.2.14) along the concentric circles of constant luminance $b(R)$, regularly divided by a sufficient number of points where we read the corresponding density and from it the transmission coefficient $T(r)$. The second integration is then achieved by summing up the values obtained from the first integration taking into account the radii of the circles and the number of points on it.

By this method we can compute [19] the density of the shadow for any individual eclipse and can also establish world tables of densities as a supplement to or refinement of the normal density tables given previously. The differences between the two are generally small (Fig. 1.2.15). The reason lies evidently in the averaging process of the integration. The effective part of the shadow terminator embraces a large amplitude of latitudes [Eq. (1.9)] and consequently the resulting density is the mean density over this interval.

1.2.17. High Absorbing Layers

In addition to the action of the Rayleigh atmosphere, we may now consider the influence of high absorbing layers and first the *ozone layer*.

The ozone distribution with height is now known from the ascent of balloons and rockets for different situations depending on the latitude and the season (Fig. 1.2.16). If we compute the total amount of ozone $O(h_0)$ along the grazing height h_0 we may proceed as for the air mass, using the integral

$$O(h_0) = \int_{h_0}^{\infty} o(h) \sec i \, dh$$

or with the variable

$$Z = \cos i$$

we get finally

$$O(h_0) = \int_0^1 o(h) N \, dZ, \quad N = L \rho^{-1}. \quad (1.55)$$

We can, of course, as many authors have done, neglect the refraction and compute directly

$$O(h_0) = \sqrt{2a} \int_{h_0}^{\infty} \frac{o(h)}{\sqrt{h - h_0}} \, dh. \quad (1.56)$$

Some results of these computations are given in the Table 1.13 [19].

For the sake of completeness, we should also discuss the influence of the *high absorbing layer* above the ozonosphere. We shall do it on two extreme models.

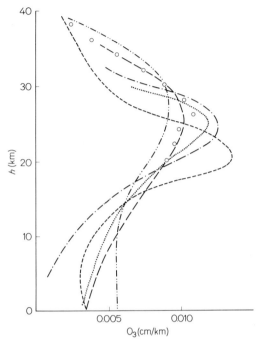

Fig. 1.2.16. Ozone repartition in altitude: — ·· — Götz at Arosa, —————— Dobson and Meetham at Tromsö, ······ Regener in Germany, — · — · — Coblentz and Stair in Washington, — — — — mean value in the Table 1.13, ∘∘ rockets 1952

Table 1.13. *Mean ozone layer*

h_0 (km)	$o(h_0)$	$O(h_0)$	h_0 (km)	$o(h_0)$	$O(h_0)$
0	3.3	32.2	25	10.3	29.0
5	4.1	32.8	30	9.1	17.4
10	5.5	33.6	35	4.6	7.1
15	7.2	33.4	40	1.5	1.2
20	9.1	32.1			

a) A *thin layer* situated between the heights h_1 and h_2 (Fig. 1.2.17). From the geometry we get for the ratio

$$\frac{G(h_0)}{G(0)} = \frac{\sqrt{h_2-h_0}-\sqrt{h_1-h_0}}{\sqrt{h_2-h_1}}. \qquad (1.57)$$

Its value shows only a small variation in the first 20 km of h_0. The rapid rise begins much higher and the maximum is reached at h_1, followed by a very rapid decrease to the upper level h_2

If the optical density in the vertical direction is B, the density corresponding to the grazing ray at $h_0 = 0$ is then

$$G(0) = 2B \sqrt{\frac{a}{2h_1}}. \qquad (1.58)$$

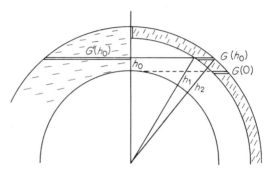

Fig. 1.2.17. High absorbing layer, left thick, right thin

b) A *thick absorbing layer* between the Earth's surface and the level at h_2. The above mentioned ratio will now be

$$\frac{G'(h_0)}{G'(0)} = \sqrt{1 - \frac{h_0}{h_2}}. \qquad (1.59)$$

It is also nearly constant up to 20 km and then decrease to the upper level at h_2. The optical density expressed by the zenithal density B is now twice as large as for the thin layer

$$G'(0) = 2B \sqrt{\frac{2a}{h_2}}. \qquad (1.60)$$

As may be seen from the above discussion, the main property of any high absorbing layer (thin, thick, ozone) is a relative constancy of absorption in the lower part of grazing rays which contrasts with the steep decrease of the extinction in the Rayleigh atmosphere.

1.2.18. Atmospheric Illumination of the Eclipsed Moon

Besides the direct solar light flux refracted and attenuated in the atmosphere, the Moon receives the scattered light from it, as was first suggested by ROUGIER and DUBOIS [20]. However, the ratio of both illuminations must be thoroughly estimated, better still calculated, before we come to any conclusion about the real influence of the atmospheric light.

Neglecting the multiple scattering, we may start [21] from the elementary principle which brings into the relation the optical density dD of an elementary volume of the scattering medium with its surface brightness db

$$db = KE\, dD \tag{1.61}$$

where E is the solar illumination and K the coefficient, depending on the nature of the scattering medium and the scattering angle. For the orthotropic scattering we have

$$K = (4\pi)^{-1}$$

and for the molecular scattering

$$K = 3(1 + \cos^2 i)(16\pi)^{-1}. \tag{1.62}$$

Let us consider the surface brightness of a diffusing medium between the points A and B (Fig. 1.2.18) in the direction of incoming solar rays,

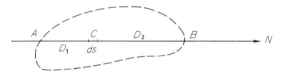

Fig. 1.2.18. Illumination by the diffusing medium

as is nearly the case for lunar eclipses. For the element ds at point C, the elementary brightness viewed from N will be

$$db = KE_0 \exp(-D_1) \exp(-D_2)\, dD \tag{1.63}$$

where D_1, D_2 are respectively the optical densities between AC and CB, and E_0 the solar illumination at the entry A into the medium. Writing for

$$D_1 + D_2 = D_0 \tag{1.64}$$

we obtain by integration

$$b = KE_0 D_0 \exp(-D_0). \tag{1.65}$$

The value of b depends on D_0 and its upper limit $\exp(-1)KE_0 = 0.368\ldots KE_0$ is attained for $D_0 = 1$.

Returning to the terrestrial atmosphere, we shall consider the brightness of its elementary ring situated between the altitudes h_0 and $h_0 + dh_0$ (Fig. 1.2.19). This ring produces the illumination

$$d\eta = b\, d\Omega \tag{1.66}$$

where the solid angle of the ring observed from the Moon is

$$d\Omega = \frac{2\pi \pi_\mathbb{C}^2}{a} dh_0. \tag{1.67}$$

We have therefore

$$d\eta = \frac{2K\pi \pi_\mathbb{C}^2}{a} E_0 D_0 \exp(-D_0) dh_0$$

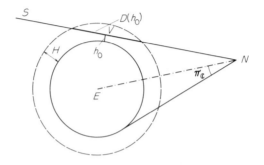

Fig. 1.2.19. Illumination by the terrestrial atmosphere

or by integration we get for the ratio η/E_0

$$\frac{\eta}{E_0} = \frac{2\pi \pi_\mathbb{C}^2 K}{a} \int_0^H D_0 \exp(-D_0) dh_0. \tag{1.68}$$

This integral may be simplified in two ways. Noting firstly that the upper limit of $D \exp -D = 0.368 \ldots$, we obtain for the upper limit of the integral

$$\frac{\eta}{E_0} < 3.63 \times 10^{-4} K \pi_\mathbb{C}^2 H_{km}. \tag{1.69}$$

Secondly, in Rayleigh atmosphere the optical density can be approximated to the exponential function

$$D_0 = D_{00} \exp - \beta h_0 \tag{1.70}$$

and the integral takes the form

$$\frac{\eta}{E_0} = \frac{2\pi \pi_\mathbb{C}^2 K}{a\beta} [1 - \exp(-D_{00})] \simeq \frac{2\pi \pi_\mathbb{C}^2 K}{a\beta}. \tag{1.71}$$

As the term depending on the wave-length $\exp(-D_{00})$ is very small, the atmospheric illumination by the Rayleigh atmosphere will be nearly independent of the colour.

These approximate evaluations may be, however, supplemented by the more exact process of numerical integration. We start from the Eq. (1.66)

$$d\eta = b(h_0) \, d\Omega$$

and we compute the integral [Eq. (1.61)]

$$b(h_0) = \int_S^N KEA \, e^{-AM_2} \, ds \tag{1.72}$$

where ρ is the air density at B, M_2 the air mass between B and N, E the solar illumination at B and A the absorption coefficient.

We divide the above integral into two parts SV and SN (Fig. 1.2.19). In the first part we consider only the extinction, neglecting the attenuation by the refraction, and the solar illumination will consequently be

$$E_1 = E_0 \, e^{-AM_1} \tag{1.73}$$

and the first part of the integral

$$b_1(h_0) = KA \int_S^V e^{-A(M_1 + M_2)} \rho \, ds = KA \, e^{-AM} \, \frac{M}{2} \tag{1.74}$$

where M is the total air mass from S to N.

In the second part of the integral, where the attenuation by the refraction cannot be neglected, we take the solar illumination from the Twilight Tables [22] and the integration may be achieved

$$b_2(h_0) = KA \int_V^N E_2 \, \rho \, e^{-AM_2} \, ds \tag{1.75}$$

by numerical integration.

With the value of $b(h_0)$ we compute the integral

$$\frac{\eta}{E} = \frac{2\pi \pi_{\mathbb{C}}}{a} \int_0^\infty [b_1(h_0) + b_2(h_0)] \, dh_0. \tag{1.76}$$

Its numerical values compared with the observed illumination in the shadow center

	Red (7000 Å)	Red (6300 Å)	Blue (4500 Å)	Blue (4600 Å)
η/E	1.9×10^{-7}		2.1×10^{-7}	
e/E		6×10^{-5}		10^{-6}

show that the atmospheric illumination by the Rayleigh atmosphere attains an observable limit only in the blue part of the spectrum. However, in a heavily polluted atmosphere, these conditions may change considerably as will be explained later (see 1.3.11).

1.2.19. Eclipse Phenomena in the Cislunar Space

SCHILLING and MOORE [23] have discussed the eclipse phenomena from the viewpoint of an observer going along the shadow axis from the Earth through the cislunar space towards the Moon. In these conditions the scattering of light is predominant.

Neglecting refraction for the time being, we have a kind of simple twilight scene (Fig. 1.2.20) analogous to that of terrestrial conditions the

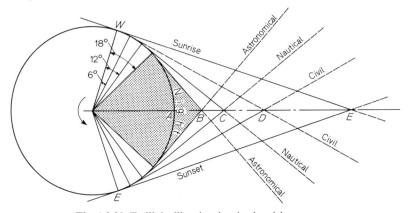

Fig. 1.2.20. Twilight illumination in the cislunar space

only difference being that the observer is placed outside the scattering medium of the terrestrial atmosphere. In his trip from A to B he is plunged into the astronomical night. Between B and C he moves through the zone of astronomical twilight, from C to D he has the nautical and from D to E the civil twilight. Finally at the vertex E the Sun rises for him.

The heights of all these points may be computed according to the formula
$$h = a[\operatorname{cosec}(U + R_\odot) - 1] \qquad (1.77)$$
where we put successively $U = 18°$, $12°$ and $6°$ for the subsequent twilight stages and $U = \omega = 1, 2°$ for the sunrise. We get following altitudes:

Stage	Night (A)	Astr. Tw. (B)	Naut. Tw (C)	Civ. Tw. (D)	Sunrise (E)
h	0	14000	23600	52100	252000 km

For the photometrical side of the problem, where the authors compute the illumination produced on the shadow axis by the luminous halo of the terrestrial atmosphere, a partial solution has been found based on a semi-empirical method that combines the refraction and multiple-scattering theory with observed brightness by KOOMEN et al. and RICHARDSON et al. [23]. Fig. 1.2.21 gives the variation of the stellar

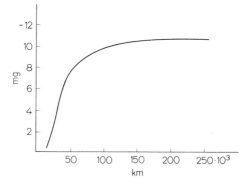

Fig. 1.2.21. Magnitude of the terrestrial atmosphere observed from the cislunar space

magnitude of the whole halo. Extrapolating this last curve to the Moon's distance we obtain with $m_\odot = -26.8$ the ratio of atmospheric and solar illumination outside the eclipse the value 3×10^{-7} instead of 2×10^{-7} obtained from Eq. (1.76) taking into account only the primary scattering.

1.2.20. Lunar Eclipses on the Moon

The eclipse of the Moon is for a lunar observer an eclipse of the Sun by the Earth surrounded by the atmosphere whose influence modifies substantially the geometrical scene of the phenomenon [1].

From a point N on the lunar surface (Fig. 1.2.22) situated at the angular distance γ from the center of the shadow, the lunar observer will see the centers of the Sun and the Earth at the angular distance

$$\gamma' = \frac{\pi_\mathrm{C}}{\pi_\mathrm{C} + \pi_\odot} \gamma \tag{1.78}$$

and their angular radii will be

$$R'_\odot = \frac{\pi_\mathrm{C}}{\pi_\mathrm{C} + \pi_\odot} R_\odot, \tag{1.79}$$

$$\psi_c = \pi_\mathrm{C}(1 + c\,\rho_c) \tag{1.80}$$

where ρ_c is the air density at the cloud level limiting the transparency of the low atmosphere. The differences between γ' and γ or $R'_\odot$ and $R_\odot$ amount to only 0.25% and can be neglected.

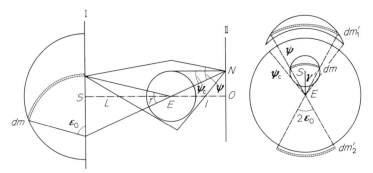

Fig. 1.2.22. Formation of the refraction image. Left: Optical schema of the imaging action. Right: Aspect of the Earth and the Sun from the Moon

The rays going from N below the angle ψ (Fig. 1.2.22) define on the solar disk an arc dm of the circle visible from the Earth below the angle

$$r = (\pi_\odot + \pi_\mathbb{C})\left(1 + \frac{h'_0}{a}\right) - \omega \tag{1.81}$$

and from the Moon the same arc will be visible as dm'_1 below the angle [Eq. (1.10)]

$$\psi = \frac{a + h'_0}{l_2} = \pi_\mathbb{C}\left(1 + \frac{h'_0}{a}\right) \tag{1.82}$$

or the difference $\Delta\psi$ giving the apparent angular distance from the Earth's limb

$$\psi - \psi_c = \Delta\psi = \pi_\mathbb{C}\left(1 - c\,\rho_c + \frac{h'_0}{a}\right). \tag{1.83}$$

Eqs. (1.80) and (1.81), giving the relation between the solar arc and its image as seen from the Moon, define some kind of optical imagery of the solar disk by the terrestrial atmosphere. The whole refraction image will be obtained by a juxtaposition of a sufficient number of arcs dm'. It may be noted that the position angles are not modified, only the radial distances from the Earth's center. In such a manner the width of the solar disk is considerably compressed by refraction. As long as the solar disk is hidden by the dark disk of the Earth the width of the solar image cannot exceed the angular width of the atmosphere, i.e. a fraction of a minute of an arc.

Fig. 1.2.23 shows some characteristic aspects of a solar eclipse as viewed from the Moon. The distances $\Delta\psi$ from the terrestrial limb are there exaggerated in order to make the form of the images more distinct.

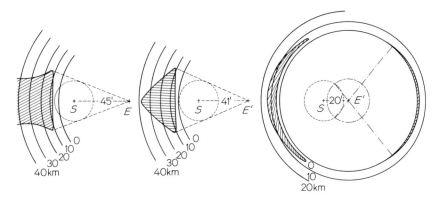

Fig. 1.2.23. Different aspects of solar images. The heights over the Earth's surface are amplified 100 times

1.2.21. Other Photometrical Theories

The photometrical theory which we have expounded stems from the principle of auxiliary shadow and has the advantage of theoretical clearness in relation to the structure of the terrestrial atmosphere. Its shadow thrown by a point source is precisely the auxiliary shadow.

On the other hand, the first theories elaborated by HEPPERGER [24] and later completed by SEELIGER [14] were built up on another basis, namely from the viewpoint of a lunar observer looking towards the Sun, as has been explained elsewhere (1.2.20).

The observer sees the elementary ring of the solar image whose surface brightness is attenuated by extinction in the ratio 10^{-AM}. Thus the illumination given by this element at the lunar point N will be according the photometrical principle

$$de = k\, b\, 10^{-AM}\, dm' \qquad (1.84)$$

where dm' is the solid angle of the element as seen from the Moon.

The total illumination will be given by the integration taken over the whole refraction image. HEPPERGER assumes $b=1$, whereas SEELIGER takes into account the limb darkening.

It seems at first sight that the attenuation by refraction has been neglected. This is not in fact the case, as this factor is implicitly comprised in the reduction of the solid angle dm' in comparison with the non-

4 Link, Eclipse Phenomena

refracted solid angle dm. We have according to the Fig. 1.2.22

$$dm' = 2\varepsilon\psi\,d\psi \qquad (1.85)$$

$$dm = 2\varepsilon r\left(\frac{L}{L+l}\right)^2 dr$$

and their ratio

$$\frac{dm}{dm'} = \frac{r\,dr}{\psi\,d\psi}\left(\frac{L}{L+l}\right)^2 \qquad (1.86)$$

gives after some easy substitutions

$$\frac{dm}{dm'} = \left[1 - \frac{\omega}{\pi_\odot + \pi_{\mathbb{C}}}\left(1 - \frac{h'_0}{a}\right)\right]\left[1 - a\frac{d\omega}{dh'_0}\frac{1}{\pi_\odot + \pi_{\mathbb{C}}}\right] \qquad (1.87)$$

the expression which is identical with our Eq. (1.35) for the attenuation by refraction.

However, SEELIGER, seems to not be conscious of the existence of this factor hidden in his theory as will be explained later (1.5.6). Seeliger's theory was adopted also by SAUSSURE [25] but, having confounded h_0

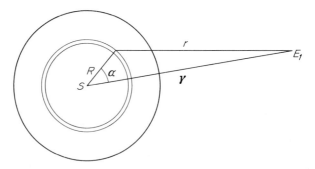

Fig. 1.2.24. Integration in solar plane according to FESENKOV

with h'_0, the author made an error of some 5–12% in the illumination by grazing rays between $h_0 = 14$ and 5 km.

Finally, we shall mention the theory of FESENKOV [26]. He also integrates in the solar plane, as we did, but he chooses for the first integration element the elementary ring described from the center of the Sun (Fig. 1.2.24). In this way he is obliged to carry out the double integration

$$c = k \iint T(r)\,b(R)\,R\,dR\,d\alpha \qquad (1.88)$$

by the numerical method, the advantage of one analytical integration having been lost by the choice of an unsuitable integration element.

In the first version of his theory, FESENKOV [26] knows only the second term in the attenuation by refraction [Eq. (1.35)] putting implicitly $\omega=0$. This approximation is permitted only on the periphery of the shadow. In the second version, FESENKOV [27] gives the complete expression with $\pi_\odot=0$ which is without any importance from the numerical viewpoint.

1.2.22. Old and Classical Theories of Refraction and Air Mass

These theories adopted by different astronomers (HEPPERGER, SEELIGER and SAUSSURE) have in common two weak points namely:

a) The aerological function $\rho=f(h)$ is expressed by an analytical formula less fitted to the actual structure of the atmosphere than to the necessity for an analytical integration of the refraction or the air mass. Many ingenious methods were used with relative success in the field of classical astrometry where the zenithal distances rarely exceed 75°. According to Oriani's theorem [28], the refraction and the air mass are merely the function of the local density of the air at the observing point, and the structure of the atmosphere along the ray enters in to the formula only as a correction term.

In the eclipse theory this is not the case and the atmospheric structure at and over the vertex of the light path has a determining influence on the refraction.

b) The air mass was computed according to Laplace's theory [28] demonstrating the proportionality between the refraction and the air mass. The two corresponding integrals

$$\omega = 2 \int_{h_0}^{\infty} c\, \rho\, \beta\, \mathrm{tg}\, i\, dh \qquad (1.89)$$

$$M = 2 \int_{h_0}^{\infty} \rho \sec i\, dh$$

can be expressed by means of the invariance theorem

$$r\,\mu \sin i = r_0\, \mu_0 \sin z \qquad (1.90)$$

as follows

$$\omega = 2 \int_{h_0}^{\infty} \frac{c\, \beta\, r_0\, \mu_0 \sin z}{r\, \mu} \rho \sec i\, dh, \qquad (1.91)$$

$$M = \frac{\operatorname{cosec} z}{\mu_0 (a+h_0)} \frac{\mu(a+h)}{c\,\beta} \omega \qquad (1.92)$$

where the variation of $\mu(a+h)$ is of secondary importance.

The proportionality assumes the constant value of

$$\frac{M\beta}{\omega} = k \tag{1.93}$$

a condition which is not satisfied as shown in the following table.

Table 1.14

h	0	5	10	15	20	30	40 km
M/ω	1.065	1.037	0.853	0.806	0.801	0.810	0.87
$\beta M/\omega$	1.290	1.109	1.114	1.254	1.245	1.260	1.33

The main reason for the inconstancy of k is the change of β with the height. FESENKOV tried to correct this change by dividing the atmosphere into three sections having different values of β but constant in each section. However, in the troposphere between 2 and 12 km, where FESENKOV adopts the constant value of β, the actual variation of it is too large to be neglected.

Let us comment once more on different refraction theories. HEPPERGER [24] and SAUSSURE [25] adopted Bessel's old theory which starts with the assumption of the aerological function

$$\rho = \rho_0 \exp - 632 \frac{h - h_0}{a} \tag{1.94}$$

and leads to the development of the following form

$$\omega = f_1(h) f_2(h) [1 + 0.413 f_1^2(h) + 0.266 f_1^4(h) + 0.196 f_1^6(h) + \cdots] \tag{1.95}$$

where $f_1(h)$ and $f_2(h)$ are the functions of the height h. It is evident that the relation (1.94), giving too small a decrease of density with height in the stratosphere, will also give there too great values of refraction.

SEELIGER used [14] the Ivory theory based on the temperature-density relation

$$\frac{T}{T_0} = 1 - k(1 - \rho) \qquad k = 2/9 \tag{1.96}$$

which gives further

$$h = 3.66(1 - \rho) - 14.74 \log \rho \tag{1.97}$$

and for the refraction SEELIGER obtained

$$\omega = 61.97' \, \rho + 5.95' \, \rho^2. \tag{1.98}$$

Fesenkov's method of computation [27] is based on the development in the series. If a general point of the light trajectory has the polar coordinates r, φ we can write for the angle i

$$i = z + \frac{f'(\varphi)}{1!} \varphi + \frac{f''(\varphi)}{2!} \varphi^2 + \frac{f'''(\varphi)}{3!} \varphi^3 + \cdots . \tag{1.99}$$

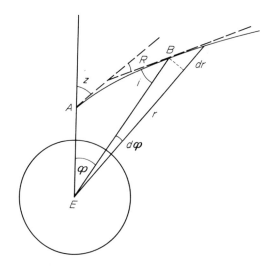

Fig. 1.2.25. Refraction theory according to FESENKOV

From the Fig. 1.2.25 we have

$$dr = r \, d\varphi \, \cotg i$$

and the invariance relation gives

$$r \mu \sin i = \text{const}$$

so that we obtain

$$f'(\varphi) = \frac{di}{d\varphi} = 1 - \frac{r}{\mu} \frac{d\mu}{dr}$$

and similarly for

$$f''(\varphi), \quad f'''(\varphi) \ldots .$$

As the refraction between the points A and B (Fig. 1.2.25) is the modification of the direction, we obtain for it

$$R = i - z + \varphi .$$

Finally we get for the refraction the series

$$R = k_1 \varphi + (k_1 + k_1^2 + k_2)\frac{\varphi^2}{2}\cotg z + [(1-k_1)(k_1 + k_1^2 + k_2)$$
$$+ 2(k_1 + 3k_1^2 + 4k_2 + k_1^3 + 2k_1 k_2)\cotg^3 z]\frac{\varphi^3}{6} + \cdots \qquad (1.100)$$

with refraction factors

$$k_1 = -\frac{r}{\mu}\frac{d\mu}{dr}, \quad k_2 = -\frac{r^2}{\mu}\frac{d^2\mu}{dr^2}, \quad k_3 = -\frac{r^3}{\mu}\frac{d^3\mu}{dr^3} \qquad (1.101)$$

which are, of course, variable with $r = a + h$. It is, therefore, necessary to divide the whole integration interval into a sufficient number of smaller subintervals and proceed in each with a new series having new refraction factors k computed for the beginning of each partial interval.

Bibliography

1. LINK, F.: Bull. astron. **8**, 77 (1933).
2. HEYDEN, F.J.: Ap. J. **118**, 412 (1954).
3. JULIUS, W.H.: Ap. J. **23**, 312 (1906); **37**, 225 (1913).
4. l.c. [1] p. 103; — ŠVESTKA Z., BAC **1**, 48 (1948); — LINK, F.: Bull Soc. Math. Phys. Tchécosl. **72**, 65 (1947).
5. ABBOT, C.G.: Ann. Astrophys. Observ. Smiths. Inst. **4**, 221 (1922).
6. LINK, F.: BAC **9**, 169 (1958).
7. CABANNES, J.: Sur la diffusion moléculaire de la lumière. Paris 1929.
8. VAN DE HULST, H.C.: Light scattering by small particles. New York 1957.
9. ABBOT, C.G.: Ann. Astrophys. Observ. Smiths. Inst. 1—4 (1900—1922).
10. HUGON, M., et F. LINK: Rev. opt. **9**, 156 (1930).
11. GUTH, VL. u. F. LINK: Met. Z. **40**, 392 (1942).
12. LINK, F., and L. NEUŽIL: Dioptric tables of the earth's atmosphere. Publ. Astr. Inst. Prague 1965, No. **50**.
13. GLEICHEN, A.: Verhandl. deut. physik. Ges. **2** (2) (1889).
14. SEELIGER, H.: Abhandl. bayer. Akad. Wiss. II. Kl. 19 (1896).
15. LINK, F.: Space Research **8**, 1069 (1968).
16. LINK, F.: J. Obs. **20**, 165 (1937).
17. RAMANATHAN R.R.: Nature **123**, 834 (1929).
18. Mc MURRAY, W.M.: Cross sections for the IGY period, US Depart. of Com. Wheather Bur., Washington D.C., 1961.
19. GUTH, VL., u. F. LINK: Z. Aph. **20**, 1 (1941); — BOUŠKA, J., et F. LINK: Compt. rend. **224**, 1483 (1947).
20. ROUGIER, G., et J. DUBOIS: Ciel et terre **60**, No. 4—6 (1944).
21. LINK, F.: Bull. astron. **15**, 143 (1950).
22. LINK, F.: Mitt. Beob. Tschech. Astr. Ges. **6**, 1 (1941).
23. SCHILLING, G.F., and R.C. MOORE: Ist lunar Int. Lab. Sympos. Athens 1965. Berlin-Heidelberg-New York: Springer 1966, p. 85.
24. HEPPERGER, J.: Sitzber. Akad. Wiss. Wien, Math. Kl. **54** (1895).
25. SAUSSURE, M.: Verhandl. naturforsch. Ges. Basel **42** (1931).
26. FESENKOV, V.: Bull. Acad. Sci. S.S.S.R. **1** (1932).
27. FESENKOV, V.: Astron. Zhur. **14**, No. 5—6 (1937); — Meteoritika **18**, 125 (1960).
28. BEMPORAD, A.: Enz. Math. Wiss, VI, **2**, 329 (1908).

1.3. Photometry of Lunar Eclipses

1.3.1. Measurements of the Shadow Density

The correct measurement of the shadow density is a very delicate problem of exact photometry over a large intensity interval. The differences in the density to be measured are generally large, of the order of $4-5$, i.e. the intensity varies in the ratio of $1:10000$ or 100000. That demands a very good linearity of the photometer scale guaranteed over such a great amplitude.

To compare our results with the theory we must create, in addition, a sufficiently narrow spectral interval, e.g. by some convenient color filter. Some blue filters have a second transmission window in the far red. When the photometric receptor is only slightly sensitive in the far red, there exists a danger of a perturbation because the red intensity of the eclipsed Moon may be some $100-1000\times$ greater than the blue in the central part of the shadow. For this reason, the measured intensity can be greater than the actual blue intensity.

But the biggest source of perturbations arises from the existence of the parasitic light during the partial phase. There we have, of course, simultaneously the penumbral crescent of the Moon $100-1000\times$ brighter than the part immersed in the umbra, where the intensity is to be measured. The very bright light of the penumbral part is diffused in the Earth's atmosphere and on the optical surfaces of the observation instrument and adds an appreciable amount of light to the feeble light of the umbra. Thus we obtain apparently a larger intensity or a lesser density than in the absence of the parasitic light. To obtain a correct value of the density we should arrange our method of measurements in such a way so as to eliminate or diminish considerably this undesirable light.

Finally we should not forget the influence of the extinction in the Earth's atmosphere between the observer and the Moon. This is, however, a danger common to all photometric measurements in astronomy, when the object changes its zenithal distance during the measurements. As the duration of a lunar eclipse may be $1-3$ hours with subsequent changes in the lunar position on the sky, the influence of the extinction must be taken into account.

1.3.2. Visual Method

All kinds of visual photometers have been used during lunar eclipses and mostly without regard to the veil of diffused light. The results have, therefore, for the above mentioned reasons little value for the photometrical theory.

DANJON's cat's eye photometer used with success by him [1], and his collaborators is almost free from the described perturbations. The principle is as follows: Two prisms (Fig. 1.3.1) P_1 and P_2 partly cover the objective O in such a way that the observer sees in the ocular two images of the Moon. By turning the whole system of prisms about the optical axis and adjusting the mutual inclination of both prisms, we can see the two lunar images just in contact in the direction to the center of the

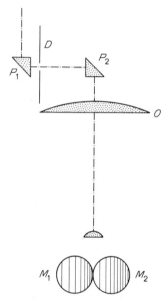

Fig. 1.3.1. Danjon's cat's-eye photometer. M_1 and M_2 the position of lunar images during measurements

shadow, i.e. in the direction of the greatest intensity variation. The cat's eye diaphragm D inserted in the light path of the prism enables us to equalise the brightness of the two parts 1 and 2 of the Moon's surface.

During the eclipse we always proceed in such a manner that the points 1–2; 2–3; 3–4 are photometrically connected (Fig. 1.3.2) and the whole light curve is consecutively constructed beginning with the penumbra, then through the umbra and ending anew with the penumbra. Outside the eclipse the whole limb of the Moon must be measured in order to eliminate the differences of the albedo. The spectral selection is realized by means of color filters.

Because of the short distance ($2R_\mathrm{C}$) which separates the two measured spots of the Moon, the influence of the extinction is very small. The influence of diffused light, as Danjon states, is without danger while both images are projected on the same background.

Danjon's method has, however, two disadvantages. The first is that we are limited to those areas of the Moon, which are situated on the end points of the lunar diameter directed towards the center of the shadow. This limitation makes it impossible to trace the isophotes of the shadow or to study the lunar luminescence (1.4) on other parts of the lunar disk than its limb. Secondly, the successive linking of measured points presuposes that the continual observations are not perturbed by cloudiness.

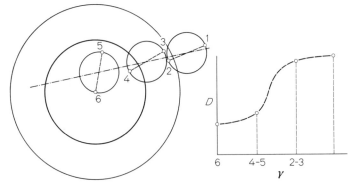

Fig. 1.3.2. Technique of measurements with the cat's-eye photometer. Left: Positions of measured points. Right: Their positions on the light curve

In spite of these circumstances Danjon's method has produced a long series of reliable observations of great value by their homogeneity which enables us to compare it with the theory (1.3.5).

1.3.3. Photographic Method

GUTH and LINK [2] used the photographic method inspired partly by Danjon's visual method. Again the penumbral and umbral parts of the Moon were compared and thus the extinction influence was eliminated. To obtain the comparable density of these very different brightnesses, a thin gelatine filter F (Fig. 1.3.3) of convenient density (~ 2) was placed close to the photographic plate. The filter has a circular aperture of the diameter corresponding to the umbra in the focal plane of the photographic camera and the image M of the Moon formed by the objective O was placed in such a way that the penumbral part of the Moon was attenuated by the gelatine filter. During the totality the Moon's image was placed naturally inside the aperture. For the photometric calibration, a photometric scale S illuminated by the constant electric bulb L was used and exposed simultaneously with the Moon. A color filter C provides the necessary spectral selection.

The influence of the diffused light must be considered and therefore the veil round the image has to be measured. At the reduction we proceeded according to Danjon's method linking progressively different parts of the umbra and the penumbra but with the possibility of measuring every spot on the Moon. In this way the isophotes of the shadow were obtained (1.3.10).

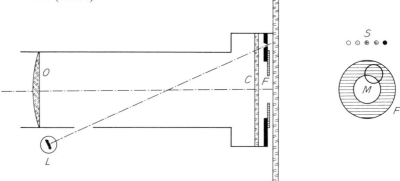

Fig. 1.3.3. Photographic eclipse photometer

1.3.4. Photoelectric Method

This very sensible method has also been used but without complete elimination of the parasitic light. WALKER and REAVES [3] have tried to attain it partly by measuring always the part of the Moon at the greatest distance from the shadow center. It is, therefore, the brightest and the influence of the diffused light is the least, if not negligible. The extinction, however, must be taken fully into account.

FRESA and CIMINO [4] compared photoelectrically two opposite parts of the lunar diameter and in this manner the extinction was eliminated but a part of the diffused light was included in the measured flux.

SANDULEAK and STOCK have photometred [5] the penumbra (December 19, 1964) by means of a refrigerated multiplier and a Weitbrecht-type integrator through a yellow filter. The field centered on a region of Marue Nubium had a diameter of 10″.

1.3.5. Comparison of the Theory with Observations

The aim of the photometric theory is its comparison with observations. The theory may have a descriptive character being built on a sufficiently large number of parameters, i.e. the structure of Rayleigh atmosphere with the given distribution of ozone and aerosols. Or the theory may have an exploratory character on the basis of the Rayleigh atmosphere only and aiming to investigate the unknown additional factors as mentioned above. We have deliberately chosen the second way and it is now time to proceed to compare the theory with observations.

Among several series of measurements the observations carried out by Danjon's method are most suitable as they are nearly free of systematic errors. Apart from some isolated comparisons, a general reduction of these measurements since 1921 has been given in the Catalogue of Lunar Eclipses II [6]. The presentation is very complete as can be seen from the following example.

Eclipse January 18–19, 1954.

A. Ephemeris.

T_1 = Entry into the penumbra	23 h 40 min UT
T_2 = Entry into the umbra	00 h 50 min
Position angle	81°
T_3 = Total eclipse begins	02 h 17 min
T_0 = Middle of the eclipse	02 h 32 min
Magnitude	1.04
T_4 = Total eclipse ends	02 h 47 min
T_5 = Exit from the umbra	04 h 14 min
Position angle	314°
T_6 = Exit from the penumbra	05 h 54 min
Lunar parallax	56′ 39″
Solar radius	16′ 15″
Radius of the umbra	40′ 33″
Radius of the penumbra	73′ 03″

B. Graphical ephemeris (Fig. 1.3.4) showing the positions of measured points by the cat's eye photometer.

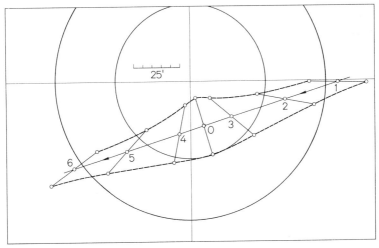

Fig. 1.3.4. Graphical ephemeris of the eclipse January 18–19, 1954. C.E.T. is used

Lunar Eclipses

Table 1.15

γ''	γ	II (5300 Å)						I (4700 Å)		
		Cr	De	C	M_0	O - C		Cr	De	C
						Cr	De			
(')	(')									
	9.3	5.65	5.65	4.00				5.96	5.96	
	10	5.46	5.24	4.06	40.4	+1.40	+1.18	5.72	5.48	5.56
	15	5.02	4.86	3.85	33.5	+1.17	+1.01	5.16	4.98	5.13
	20	4.65	4.78	3.62	27.2	+1.03	+1.16	4.96	4.97	4.68
	25	4.34	4.45	3.40	22.0	+0.94	+1.05	4.39	4.59	4.28
−10.6	30	4.20	4.24	3.19	17.4	+1.01	+1.05	4.25	4.25	3.90
− 5.6	35	4.00	3.92	2.95	12.8	+1.05	+0.97	4.05	3.92	3.44
− 4.6	36	3.98	3.86	2.89	11.7	+1.09	+0.97	4.00	3.86	3.34
− 3.6	37	3.93	3.77	2.82	10.5	+1.11	+0.95	3.93	3.74	3.23
− 2.6	38	3.84	3.68	2.75	9.2	+1.09	+0.93	3.84	3.68	3.12
− 1.6	39	3.76	3.56	2.67	7.9	+1.09	+0.89			
− 0.6	40	3.70	3.34	2.56	6.5	+1.14	+0.78			
+ 0.4	41	2.86	2.64							
+ 1.4	42	1.98	1.72							
+ 2.4	43	1.64	1.28	1.49		+0.15	−0.21			
+ 4.4	45	1.20	0.96	1.10		+0.10	−0.14			
+ 9.4	50	0.70	0.51	0.59		+0.11	−0.08			
+14.4	55	0.40	0.21	0.33		+0.07	−0.12			
+19.4	60	0.21	0.08	0.13		+0.08	−0.05			
	65	0.08	0.02							

Cr observed density during the first half of the eclipse, *De* the same during the second half, *C* mean computed density, M_0 mean air mass.

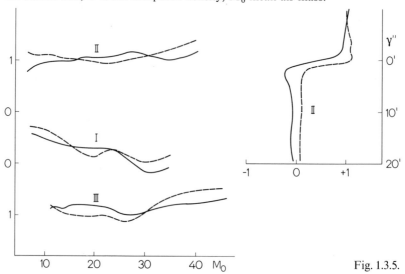

Fig. 1.3.5.

Table 1.15

M_0	O - C Cr	O - C De	III (6200 Å) Cr	III (6200 Å) De	III (6200 Å) C	M_0	O - C Cr	O - C De	IV (5800 Å) Cr	IV (5800 Å) De
			4.68	4.68					5.54	5.54
34.8	+0.16	−0.08	4.68	4.48	3.15	46.0	+1.53	+1.33	5.27	4.88
29.8	+0.03	−0.15	4.58	4.36	3.13	38.5	+1.45	+1.23	4.92	4.67
23.4	+0.28	+0.29	4.26	4.20	3.07	31.8	+1.19	+1.13	4.57	4.40
19.8	+0.11	+0.29	3.84	3.96	2.96	26.0	+0.88	+1.00	4.31	4.43
14.8	+0.35	+0.35	3.82	3.98	2.82	21.8	+1.00	+1.16	3.84	4.16
10.1	+0.61	+0.48	3.71	3.89	2.69	15.1	+1.02	+1.20	3.83	3.92
9.1	+0.66	+0.52	3.69	3.76	2.65	13.9	+1.04	+1.11	3.84	3.86
8.2	+0.70	+0.54	3.71	3.77	2.61	12.5	+1.10	+1.16	3.89	3.77
7.3	+0.72	+0.56	3.75	3.69	2.56	11.1	+1.19	+1.13	3.84	3.68

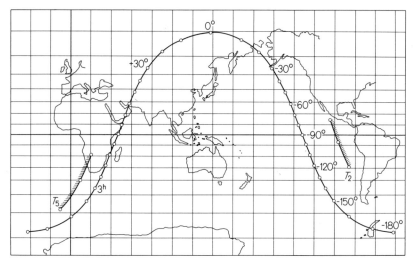

Fig. 1.3.6. Shadow terminator during the eclipse January 18−19, 1954. Successive positions: T_2 at the beginning of the eclipse, at 3h G.M.T. and T_5 at the end of the eclipse

Fig. 1.3.5. Δ-curves for the eclipse January 18−19, 1954 as a function of the mean air mass M_0 for blue (I), green (II) and red (III) light. Right: Δ-curves at the periphery of the umbra and within the penumbra as function of the angular distance γ'' from the edge of the umbra. Dashed line first half, full line second half of the eclipse

C. Visual photometry of the shadow according to DUBOIS [6]

D. The graphical representation of the differences Δ = observed − computed density (Fig. 1.3.5).

E. The chart of the shadow terminator (Fig. 1.3.6 and Table 1.1).

1.3.6. Atmospheric Ozone

As already stated the main feature of the ozone absorption in lunar eclipses is its approximate constancy up to $h_0 = 20$ km and the rapid decrease above 25 km (1.2.17). These characteristics appear clearly in the Δ-curves in the green and red parts of the spectrum where ozone has a system of Chappuis absorption bands (Fig. 1.3.7).

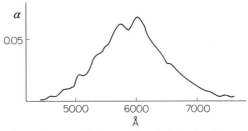

Fig. 1.3.7. Ozone absorption coefficient α (decadic) in the Chappuis absorption bands

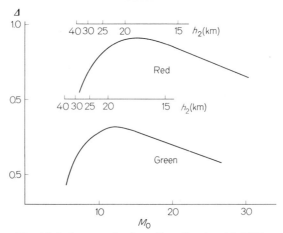

Fig. 1.3.8. Δ-curves for the eclipse October 16, 1921

A typical example is given by the eclipse of 16th October, 1921, measured by DANJON [6, 7]. We have drawn (Fig. 1.3.8) the Δ-curve and in addition to the mean air mass M_0 we represented there the scale h_2 as giving the upper limit of the illumination integral [Eq. (1.37)].

The trend of the differences Δ reflects clearly the behaviour of $O(h_0)$ function of the ozone absorption. The section of the curve between $M_0 = 25-12$ is influenced by the tropospheric layers below the 20 km level. The increase of Δ in this interval may be interpreted as a combined effect of the slight rise of $O(h_0)$ function (Table 1.13). The rapid decrease beginning at 25 km reveals the location of the upper limit of the ozone layer.

Earlier, however, CABANNES and DUFAY using their photometrical measurements of the sky light [8] placed this layer much higher at about

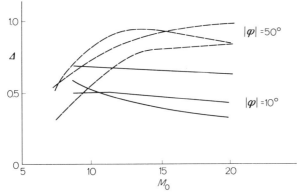

Fig. 1.3.9. Δ-curves for several eclipses at low and high latitudes

50 km, and only the first results from lunar eclipse [9] directly indicated the low level of the ozone layer, as it had been indirectly derived by means of the Umkehreffekt.

The nearly constant section of Δ-curves indicated then the maximum value of the ozone absorption. Different eclipses are placed on different parts of the shadow terminator i.e. at different geographical latitudes φ. It may be therefore expected that the behaviour of Δ-curves will change from one eclipse to another. That is what really happens if we draw the Δ-curves for some eclipses as is shown in Fig. 1.3.9. We deduce from it two points:

a) The total ozone amount increases from the equator towards the poles which is the confirmation of direct measurements.

b) The ozone layer is higher at the equator than near the poles. Dobson's measurements at Arosa (46° N) and Tromso (69° N) indicate the same phenomenon [10].

Summing up we see that even a rough discussion of the Δ-curves leads to some interesting and valid conclusions concerning the structure of the ozone layer.

PAETZOLD [11] carried out a more detailed investigations of the ozone layer based on the Eq. (1.56)

$$O(h_0)=\sqrt{2a}\int_{h_0}^{\infty}\frac{o(h)}{\sqrt{h-h_0}}\,dh.$$

From the eclipse measurements we know where the function $O(h_0)$ and the distribution function $o(h)$ is to be determined. This may be done by the solution of the integral equation of the Abel type

$$o(h)=\frac{1}{\pi\sqrt{2a}}\int_{h}^{\infty}\frac{dO(h_0)/dh_0}{\sqrt{h_0-h}}\,dh_0. \qquad (1.102)$$

The difficulty here arises from the fact that eclipse results do not give the proper value of $O(h_0)$ for a specific ray of grazing altitude h_0 but for some cone of rays illuminating the measured point in the shadow. In other words, instead of $O(h_0)$ we arrived at some mean value $O^*(h_0)$. Hence PAETZOLD substitutes for h_0 the new altitude h_0^* for which the air

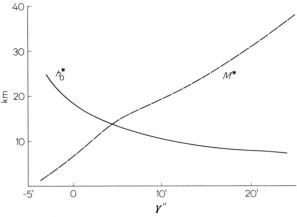

Fig. 1.3.10. Reference height (h_0^x) and the reference air mass (M^x) as a function of γ'' for 6000 Å according to PAETZOLD

mass $M^*(h_0)$ is equal to the mean air mass M_0 of the beam computed by Eq. (1.38). These quantities, called by PAETZOLD reference height $h_0^*(\gamma'')$ and reference air mass $M^*(\gamma'')$, can be drawn as the function of the angular distance γ'' from the edge of the geometrical umbra (Fig. 1.3.10) and consequently used to obtain the function $O(h_0)$ necessary for the solution of Eq. (1.102).

Paetzold's method was used for several eclipses [11]. Particularly interesting are his results relating to five eclipses (Table 1.16) similar to those in Fig. 1.3.9. Our conclusions given above a) and b) based on the

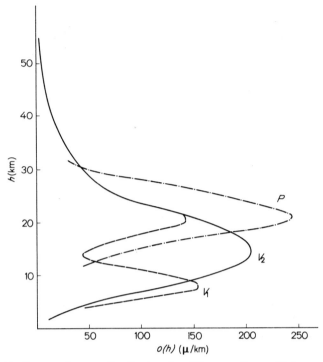

Fig. 1.3.11. Curves of ozone distribution for different eclipses: V_1 for equatorial, V_2 for polar region according to VIGROUX (eclipse January 29, 1953), P according to PAETZOLD (October 16, 1921)

Table 1.16

				φ	$O(0)$ (cm)	h_{max} (km)
1	1932	IX	14	57°S	0.36	18
2	1921	X	16	45°S	0.33	19
3	1943	II	20	49°N	0.33	16
4	1932	IX	14	0°	0.25	25
5	1942	III	2	9°S	0.22	18
6	1943	VIII	15	19°S	0.20	23

features of Δ curves are confirmed by the form of the distribution function $o(h)$ in Fig. 1.3.11 and the numbers of the Table 1.16.

VIGROUX [12] used a similar method to discuss the eclipse on January 29, 1953. Two parts of the umbra were examined. The first lying near the equator gave a double peak distribution (Fig. 1.3.11) whereas the second part in septentrional latitudes produced a simple curve.

5 Link, Eclipse Phenomena

1.3.7. High Absorbing Layer

At the present time, when the distribution function $o(h)$ of the atmospheric ozone is known under different conditions, the whole procedure may be reversed in order to detect any possible differences $\Delta' = O - C$ (R. atm. + ozone) as we did previously for the differences $\Delta = O - C$ (R. atm. only). From the known function $o(h)$, and from the absorption coefficients measured in the laboratory, we compute the ozone absorption during the eclipse and include it in the theory. For the computation of the horizontal ozone amount $O(h_0)$ we may use either the formula neglecting the refraction [Eq. (1.56)] or we can take it into account especially in lower parts of the atmosphere by using Eq.(1.55).

From these values of $O(h_0)$ we can compute the approximate constant absorption in different spectral regions using the absorption coefficients α according to VIGROUX (Fig. 1.3.7) by the formula

$$\Delta'' = \alpha \, O(0-20) \tag{1.103}$$

where $O(0-20)$ is the mean value of $O(h_0)$ in the range $0-20$ km.

λ	4600	5400	5800	6200	6800	7000	7600 Å
Δ''	0.034	0.33	0.53	0.43	0.15	0.086	0.017

However, some tolerance must be admitted as there are annual and geographic variations but their amplitude will hardly exceed 100%, as can be seen from terrestrial determinations of the vertical ozone content.

As matter of fact, we find in most cases positive residuals which cannot be explained, even with the maximal admissible ozone amount. This density excess may be visualised for instance by drawing the dependence of Δ' upon the mean air mass M_0 or the upper limit h_2 of the illumination integral [Eq. (1.37)] and comparing this curve with that for ozone absorption Δ'' (Fig. 1.3.12). There remains between the two curves difference too large to be explained by an unusually large amount of ozone and displaying an approximate constancy above the 25 km level where the ozone absorption rapidly falls. Another example of this kind is given by the eclipse on January 19, 1954 with the equatorial location of the shadow terminator (Fig. 1.3.6) With the small amount of ozone existing at low latitude, it is almost impossible to explain the large value of Δ especially its constancy between the air mass from $M_0 = 40$ to about $M_0 = 8$. These lunar eclipse phenomena, among others [13], lead us to consider the hypothesis of a high absorption layer in the Earth's atmosphere at about 100 km level.

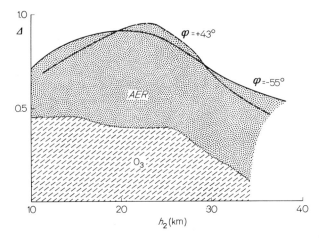

Fig. 1.3.12. Δ-curves for two eclipses compared with computed curve giving the ozone absorption

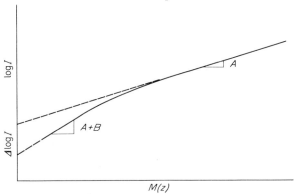

Fig. 1.3.13. Bouguer's graph

Such a hypothesis is not after all entirely new in aeronomy, as it can be traced back to the end of the last century, when MÜLLER presented his measurements of the atmospheric transmission at Potsdam [13] and at the summit (2.5 km) of Säntis [13]. These determinations were performed by means of Bouguer's method going near to the horizon and gave curves (Fig. 1.3.13) instead of the straight lines required by the simple theory. The curvature between two straight parts may be explained by the hypothesis of the high absorbing layer according to the proposition of HAUSDORFF [14] made as early as 1895.

Later, similar results were obtained under better observing conditions by BAUER and DANJON [15] on Mt. Blanc and LINK [13] on Pic-du-Midi. In addition a thorough discussion of some other observational series revealed [13] similar indications. However, the method is a very delicate operation and assumes the absence of tropospheric pollution, which is rarely the case even in high mountain regions.

Let us outline a simple theory of the Bouguer's graph. As already stated, at small zenithal distances the path in the atmosphere i.e. the air mass or the path in all its layers can be expressed simply by sec z and this part ($fz<75°$; $M<4$) of Bouguer's graph is, therefore, a straight line with the equation

$$\log I = \log I_0 - (A+B+C)\sec z. \qquad (1.104)$$

In general the path in a thin layer located at the altitude h_1 is given by the equation

$$G(z) = \left(\cos^2 z + \frac{2h_1}{a}\sin^2 z\right)^{-\frac{1}{2}} \qquad (1.105)$$

and the path in a thick layer reaching the level h is given by

$$H(z) = -\frac{a}{h_2}\cos z + \left[\left(\frac{a}{h_2}\cos z\right)^2 + \frac{2a}{h_2}\right]^{\frac{1}{2}}. \qquad (1.106)$$

Thus the equation of the Bouguer's curve will be in general case

$$\log I = \log I_0 - AM(z) - BG(z) - CH(z) \qquad (1.107)$$

where A, B, C are the optical densities in zenith of the Rayleigh atmosphere, of the thin high absorbing layer and of the low layer. Some numerical values of $M(z), G(z)$ and $H(z)$ are represented in Fig. 1.3.14. In high mountains we can neglect the low polluted layer ($C=0$). As we can see from Fig. 1.3.14, near the horizon the path $G(z)$ is practically constant and attains at horizon its maximum

$$G(90) = \sqrt{\frac{a}{2h_1}}. \qquad (1.108)$$

The equation of Bouguer's curve will be

$$\log I = \log I_0 - AM(z) - BG(90) \qquad (1.109)$$

i.e. a straight line having slope A in comparison with $A+B$ near the zenith. In this manner not only can the form of Bouguer's curve be explained but at the same time a method offers itself for the determination of the layer parameters h and B:

i) The difference between the two slopes of both linear parts of Bouguer's curve gives the optical density B of the layer in zenith.

ii) The segment on the y-axis determined by the extrapolation of both linear parts is equal to $\Delta \log I = BG(90)$ and with Eq. (1.108) we may compute the height h of the layer.

The presence of haze in the lower atmosphere, i.e. when $C \neq 0$, can almost compensate the diminution of the slope near the horizon due to the constancy of $G(z)$ while the path $H(z)$ in the haze rises very rapidly near the horizon (Fig. 1.3.14).

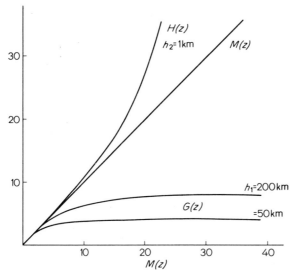

Fig. 1.3.14. Paths in different layers as a function of the air mass. $H(z)$-haze layer, $G(z)$-thin high absorbing layer

Nevertheless the results [13] seem to indicate for $B = 0.015 - 0.045$ and $h = 100$ km or little more.

Extensive observational material has been provided by the work of ABBOT and colleagues [16] as a by-product of the solar constant determinations. Only the first part of the Bouguer curve has been regularly observed, giving for its slope the sum $A + B + C$. In the Rayleigh atmosphere according to Eq. (1.28) we should have

$$\log A_1 = \text{const} + \log \frac{(\mu - 1)^2}{\lambda^4} = \text{const} + \log f(\lambda) \qquad (1.110)$$

i.e. in the logarithmic form the straight line inclined by 45°. In a polluted atmosphere the observed curves give the differences $O - C$ which are almost independent of the wavelength and on clear days also of the altitude of the observing station (Fig. 1.3.15) [19]. Therefore, the absorption B can form a part of the differences $O - C$ which are the upper limit of it. In this manner we may conclude that the maximal value of $B = 0.004 - 0.01$. Similar ideas have been expressed by LINKE [17] who was first to draw the above diagram and by GÖTZ [18].

During lunar eclipses, the hight absorbing layer manifests itself by the nearly constant shift between the observed and computed curves, the influence of the ozone having been removed (Fig. 1.3.12). As far as the real structure of the layer is concerned, the eclipse, i.e. absorption, measurements cannot decide anything for sure about it. We may discuss two extreme cases, one of the thin layer at level h and another of thick

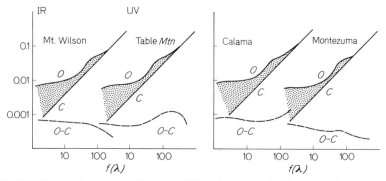

Fig. 1.3.15. Linke's diagram. Observed (O) and computed (C) values of the optical density in zenith. Below the differences $O - C$

homogeneous layer between the level h and the Earth's surface. Both models have similar behaviour, i.e. nearly constant value at large intervals of grazing heights h_0, but its value is twice as great for the thick than for the thin layer [Eq. (1.58)–(1.60)] if the zenith absorptions are assumed to be unity.

The distinction between both cases — thin or thick layer — is possible on the basis of twilight phenomena. During twilight different layers of the atmosphere enter successively in the shadow and the solar scattered light consequently diminishes. Between the optical density B of a layer at the zenith, its luminance b due to the scattered solar light and $b_\odot$ the mean luminance of the solar disk, the following relation holds [20]

$$b = 1.2 \times 10^{-5} B b_\odot \qquad (1.111)$$

if orthotropic scattering may be assumed.

At the moment during during twilight when the layers above 30 km are still illuminated, the measurements give

$$b = 10^{-9} b_\odot .$$

On the other hand, if the absorbing layers of $B=0.005$ were all above this level its luminance would be

$$b = 6 \times 10^{-8} b_\odot \qquad (1.112)$$

or about 60 times larger than observed. This implies that the absorbing layer is thick with its major part in the upper stratosphere and partly in the mesosphere. Recently the measurements of daylight scattering by VASSY [21] and RÖSSLER [22] from rockets seem to corroborate this concept. The constituents of such a layer may be aerosols of different origins. — meteoritic, terrestrial or perhaps formed *in situ* by some chemical reaction, as JUNGE suggested some years ago [23].

1.3.8. The Behaviour of Meteoritic Particles in the Atmosphere

Meteoritic matter enters the atmosphere at its cosmic velocity and is stopped by the resistance of the air at a level between 100 and 200 km without any great modification of its structure if the cosmic velocity is less than to 60 km/sec and the diameter of the particle less than 10^{-2} cm. This is the case of so-called *micrometeorites*. Larger bodies, with the exception of bolides, are dispersed approximately at the same heights. At all events the meteoritic particles begin to fall from the level of accretion towards the Earth's surface at the velocity determined by the resistance of the atmosphere.

Between the level of accretion lying not far from 100 km and the Earth's surface pressure and density changes are in the ratio of $1:10^{-6}$ and it is, therefore, evident that the law governing the fall cannot be evaluated by a simple expression. In general, during the fall at each point of the trajectory, the weight P of the particle should be balanced by the drag force of the air R. The general expression for this condition is [24]

$$P = \frac{4}{3} \pi a^3 g \delta = R = \frac{6\pi \mu a v}{1 + \frac{l}{a}\left[A + B \exp\left(-c\frac{a}{l}\right)\right]} \qquad (1.113)$$

where l the free path of molecules
 δ the density and
 a the radius of the particle assumed to be spherical
 v its velocity
 μ the coefficient of the viscosity of the air
 g the acceleration of gravity
 A, B and c constants depending upon the nature of particles.

This relation was experimentally verified by MILLIKAN [25] in the interval $0.5 < l/a < 134$.

The sum $A+B$ can be determined theoretically or by experiment. EPSTEIN [49] showed theoretically that its value depends upon the mechanism of separation which operates during the collision of the particles with the molecules. He obtained namely:

$A+B$	Separation
1.575	Specular reflection
1.091	Reflexion with conservation of velocity
1.093	Reflexion with accommodation of temperature for the non-ductive particle
1.131	As above but for the conductive particle

The experimental value of $A+B$ for oil droplets falling in air found by MILLIKAN is $A+B=1.154$, signifying a slight tendency of some 10% of molecules for the specular reflexion.

Davies' discussion [26] of all the material available in 1945 arrived at the following values

$$A=0.882, \quad B=0.281, \quad c=1.57$$

and the term

$$A+B \exp\left(-c\frac{a}{l}\right) \tag{1.114}$$

varies practically to the limits from 0.882 to 1.163.

For the free path in the air we have

$$l=\frac{\mu_0}{\rho'}\sqrt{\frac{\pi M}{kT}} = 11.85 \times 10^{-6} \frac{T}{T+113} \frac{1}{\rho'} \tag{1.115}$$

and for the coefficient of viscosity

$$\mu = \mu_0 \frac{T^{\frac{3}{2}}}{T+C} = 14.67 \times 10^{-6} \frac{T^{\frac{3}{2}}}{T+113}. \tag{1.116}$$

The general formula for the velocity becomes according to

$$v = \frac{2}{9}\frac{a^2 g \delta}{\mu}\left[1+\frac{l}{a}\left(0.882+0.281\exp\left(-1.57\frac{a}{l}\right)\right)\right]. \tag{1.117}$$

In the upper atmosphere where l/a is great we have simply

$$v_1 = 2.10 \times 10^{-2} \, a \, \delta \, \frac{g}{\rho\sqrt{T}} \tag{1.118}$$

or with $g=955$ cm/sec^2 and $T=230°$ K

$$v_1 = 13 \frac{a\delta}{\rho} \tag{1.119}$$

which is the velocity in the Millikan regime. In other words the velocity is there indirectly proportional to the air density ρ.

On the other hand, in the lower atmosphere, our formula is reduced to that of STOKES

$$v_2 = \frac{2}{9} \frac{a^2 \delta g}{\mu} \qquad (1.120)$$

or with $g = 975$ cm/sec^2 and $\mu = 160 \times 10^{-6}$ we get

$$v_2 = 1.3 \times 10^6 \, a \, \delta. \qquad (1.121)$$

The velocity in the Stokes regime is hence constant.

The duration of fall between the level h_2 and h_1, can be evaluated numerically from the velocity v with the aid of

$$t = \int_{h_1}^{h_2} \frac{dh}{v}. \qquad (1.122\,a)$$

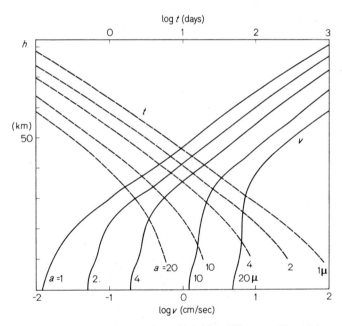

Fig. 1.3.16. The speed v and the duration of fall t for different radii a of the particles ($\delta = 1$)

In Fig. 1.3.16 the values of v and t are given for different radii of the particle having the density $\delta = 1$. For other densities the curves should be shifted by $\log \delta$ towards the right for v and towards the left for t.

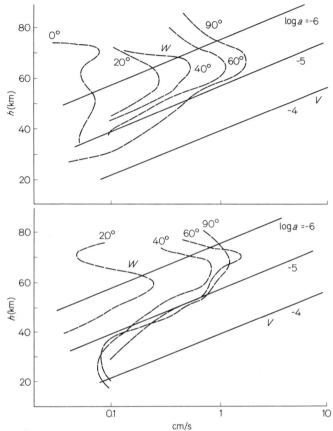

Fig. 1.3.17. Fall of particles ($\delta = 3$) in the atmosphere with vertical motions at different latitudes

Our theory and its numerical results are valid only in a quiet atmosphere without any vertical motion. In the actual atmosphere with the vertical motion having the velocity w (positive for descending current), the velocity of fall will be modified according to

$$V = v + w. \qquad (1.122\,\text{b})$$

Our knowledge of vertical motions in the atmosphere is limited. Some results by MURGATROYD [27] are represented in Fig. 1.3.17 giving the vertical motions in several latitudes for summer and winter conditions. We can see that the particles $a = 10^{-4}$ cm and $\delta = 3$ are practically unperturbed during their fall in the upper atmosphere. On the other hand, smaller particles may be stopped everywhere at the intersection of both curves v and w where $v = -w$. Theoretically we should meet in these regions an accumulation of particles falling from above but they will be removed by the lateral outflow of the air. On the average over the Earth the ascending currents must be compensated by the descending current and the model of the quiet atmosphere fits well with these conditions.

Returning to this model, let us consider what happens when the rate of meteoritic accretion is m_0 g/sec × cm² at some high level in the atmosphere. The corresponding number of particles will be

$$n_0 = \frac{3}{4} \frac{m_0}{\pi \delta a^3}. \tag{1.123}$$

During their fall in the steady state each horizontal cm² is traversed by n_0 particles per second. Between two such squares distant by v there will be n_0 particles and the concentration will be

$$n = \frac{n_0}{v} = \frac{3}{4\pi} \frac{m_0}{\delta a^3 v} = 0.24 \frac{m_0}{\delta a^3 v} \tag{1.124}$$

i.e. indirectly proportional to the velocity.

In the Millikan region we have

$$n_M = 2.3 \times 10^{-2} \frac{\rho}{a^4 \delta^2} m_0 \tag{1.125}$$

hence the concentration of particles is proportional to the air density. In the Stokes region it will be constant

$$n_S = 1.96 \times 10^{-7} \frac{m_0}{\delta^2 a^3} \tag{1.126}$$

The total number N of particles in the vertical column is given by the integral

$$N = \int_{h_0}^{h_1} n \, dh = n_0 \int_{h_0}^{h_1} dt = n_0 t \tag{1.127a}$$

i.e. equal to the number of particles accreted in the upper atmosphere during the fall t. This is indeed a natural consequence of the steady state we have assumed. Likewise for total mass in the vertical column we get

$$M = m_0 t. \tag{1.127b}$$

The presence of meteoritic particles in the atmosphere is accompanied by extinction. The optical density of the column having the section of 1 cm² and containing M grams of meteoritic matter is according to GREENSTEIN [28]

$$\varDelta = k(a) M \tag{1.128}$$

where the function $k(a)$ is to be calculated on the basis of the theory. For current meteoritic matter and in the photographic region of the spectrum, GREENSTEIN gives the following values of $f(a)$. The values marked with an asterisk may be a little low in the visual region, elsewhere the extinction is nearly neutral.

Table 1.17

a	10^{-7}	10^{-6}	10^{-5}	10^{-4}	10^{-3} cm
Iron	4.2*	4.2*	4.1*	3.0	3.0
Silicates	−1.7*	+1.4*	4.2*	3.3	2.3

The above Table gives $\log_{10} k(a)$.

Combining the Eqs. (1.127b) and (1.128) we get for the optical density in zenith [24]

$$\varDelta = k(a) m_0 t \tag{1.129}$$

This formula describes fully together with Table 1.17 the optical behavior of monodispersive (a=const) meteoritic matter in the atmosphere. It may be visualized by means of our diagram in Fig. 1.3.18 computed as to give the optical density in zenith $\varDelta = 0{,}004$ for several rates of accretion between the limits 10^{-16} and 10^{-12} g/cm² × sec. To

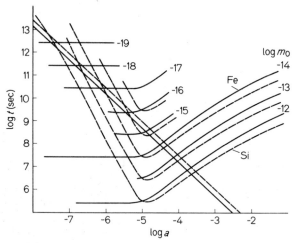

Fig. 1.3.18. Regime of the high absorbing layer with the optical density in zenith $B = 0.004$. The curves for iron (Fe) and silicates (Si) give the time of fall t (sec) as a function of the accretion m_0 and the radius a of the particle. The straight lines give the time of fall computed according to the theory [Eq. (1.122a)]

illustrate it, the particle of ferrous composition of $a = 10^{-4}$ cm requires the duration of fall $T = 4 \times 10^6$ sec with the accretion $m_0 = 10^{-12}$ or $T = 4 \times 10^7$ sec with the accretion $m_0 = 10^{-13}$. In addition we give there the duration of fall according to Eq. (1.122a) represented by two lines both meteoritic compositions.

For the duration of fall t we obtained some information in different branches. It has been shown by ŠVESTKA [29] that the luminosity of lunar eclipses expressed in Danjon's scale suffers a perceptible decline just after the maximum of meteoric showers (Fig. 1.13.19) which lasts

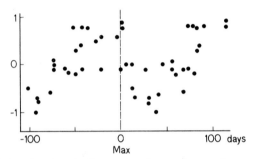

Fig. 1.3.19. Luminosity of eclipses after the maximum of meteoric activity

1—3 months before the complete recovery of the luminosity of the normal value. BOWEN [30] and his collaborators showed that the heavy rains and the number of condensation nuclei rises suddenly 30 days after the maxima of various meteoric showers. All this should mean that the great bulk of meteoritic particles needs at least one month (2×10^7 sec) to descend from the accretion level to the Earth's surface.

All these circumstances give us a rough idea about the regime of meteoritic aerosols in the Earth's atmosphere which should be confronted with direct sampling by rockets or satellites.

1.3.9. Tropospheric Influences

The principal aim of the photometry of lunar eclipses as already stated is the exploration of the upper atmosphere. Nevertheless there remain some interesting aspects in relation to the troposphere ($h_0 < 10$ km). As can be seen in the Fig. 1.2.8 the structure of tropospheric layers has a preponderant influence upon the illumination in the central parts of the umbra for $\gamma < 20'$. By examining these curves we must consider not only the altitude scale but also the angular extent of the effective terminator of the shadow on the Earth's surface. For illustration at $\gamma = 20'$ about half the illumination is due to the ray passing under $h_0 < 9$ km but the extent of the terminator is of $106°$. In this case the local incidents in the troposphere cannot have any appreciable influence on the illumination but only very large meteorological perturbations' some extended mountain chains etc. may intervene. For $\gamma \leq 16'$ the whole terminator of 360° of extent is effective and thus only the perturbations of planetary extent such as volcanic pollution could play some role in central parts of the umbra.

1.3.10. Meteorological Perturbations

Meteorological phenomena are essentially of tropospheric location and their visible manifestations — different cloud systems attaining the tropopause (11–18 km) — can perturb the normal course of lunar eclipses. We can proceed to investigate these problems in two different ways.

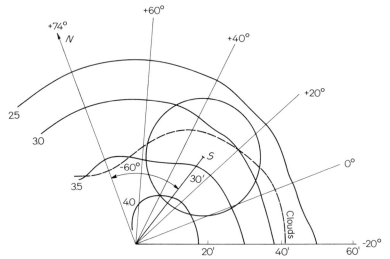

Fig. 1.3.20. Isophotes of the auxiliary shadow for the eclipse November 7, 1938

The first, is the mean state of cloudiness over the whole Earth. As it follows from the form of isophotes of the auxiliary shadow (Fig. 1.3.20), their flattening favours the solar illumination in the polar parts of the shadow. This effect is considerably enhanced by the fact that the level of highest clouds (tropopause) is also flattened in the same direction.

Cloud level at different latitudes

Latitude	0°	20°	40°	60°	80°
Ci-Str height	12.5	11	9	7	6 km

The computation of the shadow density at $\gamma = 25'$ for the further given eclipse is shown in the following numbers

Density excess over the normal densities

Position angle	0° (pole)	40°	90° (equator)
Density excess	+0.07	+0.16	+1.05

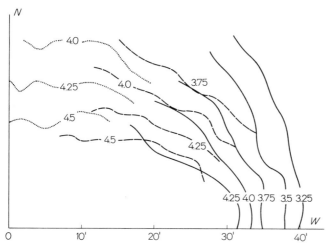

Fig. 1.3.21. Isophote of the shadow measured during the eclipse November 7, 1938

At this distance ($\gamma = 25'$) the influence of the cloud level ascending towards the equator is very pronounced. This is, of course, the maximum effect as we assume an uninterrupted belt of clouds along the shadow terminator. In reality these high clouds may be present only over a particular synoptic situation of limited extent in space and time.

In the second way, we analyse the actual synoptic situation along the terminator for a given eclipse on the basis of meteorological charts and aerological soundings. The first attempt of this kind was made for the eclipse on November 7, 1938 whose measurements [31] have exhibited very conspicuous anomalies of isophotes (Fig. 1.3.21). This behaviour consists of

i) *flattening of isophotes* growing towards the centre of the shadow,

ii) *progressive darkening* at a given point of the shadow with time.

Therefore, a meteorological analysis along the effective terminator was made by SEKERA [32] using the meteorological charts of the U.S.

Weather Bureau and the Hamburger Seewarte for the night of the eclipse. The meteorological profiles at three different moments showed (Fig.1.3.22) the cloudy and perturbed weather from the first to the last profile in the sense of growing cloudiness along the terminator. In this way the above mentioned behaviour of the isophotes is fully explained.

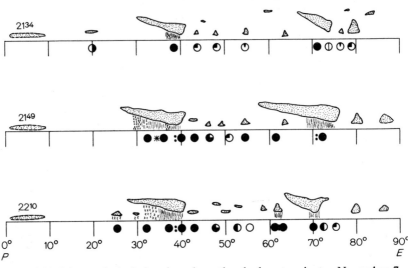

Fig. 1.3.22. Meteorological situation along the shadow terminator November 7, 1938. The circles indicate the cloudiness

It would be very interesting to undertake this analysis also for other eclipses measured in the past either on the basis of classical meteorological charts or, more recently, by using the meteorological satellites.

1.3.11. Atmospheric Pollution of Planetary Extent

The disappearance of the eclipsed Moon or, a very dark eclipse on the limit of the visibility has been only rarely recorded as is shown in the following table [33] on p. 81.

Theoretically, to the reduction of the illumination to zero in the center of the shadow at $\gamma=0$ the corresponding illumination in the auxiliary shadow must remain negligible (~ 0) to the angular distance $r=R_\odot=16'$. This requires an almost total opacity of the atmosphere up to the altitudes of 4 km ($\pi_{\mathbb{C}}=61'$) or 6 km ($\pi_{\mathbb{C}}=54'$) round the Earth at the length of 40,000 km. If the cloudiness can reach at some parts of the terminator the required altitudes 4–6 km, it is nearly impossible to assess it for the whole terminator of 40,000 km. As another factor of this kind we may consider the general atmospheric pollution by volcanic dust as a consequence of some very great volcanic explosions.

Table 1.18

No.	Date	Degree	Observers	Observations and volcanic explosions
1	1601 XII 9	11	KEPLER	Eclipsed part invisible
2	1620 VI 15	18	KEPLER	Moon invisible, stars visible
3	1620 XII 9	19	CYSAT at Ingolstadt	Moon invisible
4	1642 IV 14	19	many observers	Very dark or invisible. Explosion Avoe 1741 I, 1 km^3 ashes
5	1761 V 18	18	many observers	Very dark or invisible. Explosion Jorullo 1759 IX, 28?
6	1816 VI 16	15	LEE in London, EULE in Dresden	Moon invisible. Explosion Tambora 1815, spring, 150 km^3 ashes!
7	1884 X 4	18	many observers	Very dark and colorless. Explosion Krakatoa 1883 VIII, 18 km^3 ashes
8	1902 X 16	18	BARNARD and others	Very dark. Explosions 1902 V. Mt. Pelée and St. Vincent, 1 km^3; Santa Maria 1902 X, 5 km^3
9	1903 IV 11	12	many observers	Very dark. Explosions as above
10	1913 III 22	19	many observers	Very dark and colorless. Explosion Mt. Katmai 1912 VI–X, 21 km^3
11	1913 IX 15	17	many observers	
12	1963 XII 30		measurements	Very dark and colorless. Explosion Mt. Agung, Bali 1963 III, 17

DUFOR [34] and FLAMMARION [35] have already expressed this idea, but only recently this problem has been discussed on the basis of numerical and observational data [33].

The study of SYMONS [36] after the Krakatoa explosion in 1883 has shown the propagation of ejected volcanic dust by the general atmospheric circulation not only round the Earth in the geographical zone of the volcano but also to all geographic latitudes. The zonal atmospheric circulation is composed of three currents. The first on the equator is directed towards the west, the second in the middle latitudes towards the east and in the polar regions the third again towards the west. This repartition goes up to 25 km and the velocity of the winds is superior to 40 km/h so that a month is necessary to complete a trip round the Earth on the equator. In the meridional circulation we have three cells on every hemisphere with ascending currents at the equator and at the parallels of 60°, and descending currents at 30° and on both poles. In these conditions a worldwide spread of volcanic dust can arrive within a period of some months. As the height of ejected material can attain some tens to a hundred kilometers and the fall of fine particles in the quiet atmosphere takes several weeks, the clear up of the actual turbulent atmosphere can last several months or a year. This has actually been observed after some great explosions.

It remains finally to discuss whether the quantity of ejected dust is sufficient to produce the required opacity of the atmosphere [33]. We may assume that for a huge explosion with the total quantity of ejected dust attaining several km^3 a minimum quantity of fine dust equal to 0.1 km^3 or $M = 2 \times 10^{14}$ g will remain in the atmosphere for a number of months. The polluted layer can be 20 km high over half the Earth's surface and, therefore, the spatial concentration is of $\sigma = 4 \times 10^{-11}$ g/cm^3.

The optical density of the column having the section of 1 cm^2 and containing the mass m of fine particles is according to GREENSTEIN [28]

$$D_V = f(a) m \tag{1.130}$$

where $F(a)$ is a function of the radius a and the composition of the particle. If $a = 10^{-4}$ cm, as is indicated from the shape of Bishop's rings, we get for silicates $f(a) = 10^3$. The length of the column s in the atmosphere can be calculated from its minimal altitude h_0 or can be found in the tables [50] so that the resulting optical density D_V

$$D_V = f(a) m = f(a) s = 4 \times 10^{-8} s \tag{1.131}$$

is compared with the density in Rayleigh atmosphere D_R for green part of the spectrum.

Table 1.19

h_0 (km)	s (km)	m (g)	D_V	D_R
0	1100	$4{,}4 \cdot 10^{-3}$	4.4	—
5	920	$3{,}6 \cdot 10^{-3}$	3.6	4.1
10	740	$3{,}0 \cdot 10^{-3}$	3.0	3.2
15	514	$2{,}0 \cdot 10^{-3}$	2.0	2.7

The comparison shows that the volcanic pollution may be of the same importance as the Rayleigh scattering.

HANSEN and MATSUSHIMA [37] examined the volcanic problem again on the occasion of the Mt. Agung (Bali) explosion on March 17, 1963. Large quantities of volcanic dust were ejected to great heights in the stratosphere. Optical effects have been reported from both the northern [38] and southern [39] hemispheres where volcanic dust was also collected at a height of 20 km using the probes carried by U-2 aircraft [40].

Three eclipses, all very dark, occurred after this event and have been photometrically measured. In addition for the same period 1963–1964 we have at our disposal the astronomical determinations of extinction coefficients used in stellar photometry. Thus the complex analysis promises to be fruitful.

HANSEN and MATSUSHIMA [37] computed at first the theoretical density of the shadow on the basis of our theory (1.2) using the atmospheric structure according to U.S. Standard Atmosphere 1962. The resulting structure of the auxiliary shadow, therefore, differs slightly from our computations. To take into account the ozone absorption, the authors made use of Green's empirical formula [41] giving the vertical distribution of ozone

$$o(h) = \frac{P_1}{P_2} \frac{\exp P(h)}{[1 + \exp P(h)]^2}, \quad P(h) = \frac{h - P_3}{P_2} \quad (1.132)$$

where P_1, P_2, P_3 are the adjusted parameters. Two models have been used for two different vertical ozone contents: 0.26 and 0.36 cm.

For the dust extinction the hypothesis of a uniform layer extending from the Earth's surface to altitude H varying from 10 to 25 km was assumed and its transmission coefficient was computed with

$$T(h_0) = 10^{-\beta L(h_0)} \quad (1.133)$$

where $L(h_0)$ is the linear path in the dust layer and β the vertical extinction coefficient independent of the wave length.

To compute the illumination by the scattered light the equation similar to [Eq. (1.71)] was used

$$\frac{3 \pi_{\mathbb{C}}^2}{4 R_\odot} \int_{H_1}^{H_2} T(h_0) \beta L(h_0) dh_0 \quad (1.134)$$

where H_1 and H_2 are both limits of the dust layer.

It is clear that we can obtain by the convenient choice of the dust layer parameters a sufficiently low direct illumination which is selective and a relatively high illumination by scattering which is neutral. In other words in the central parts of the shadow the direct illumination may be supplanted by the scattered light. The authors succeeded in doing this by putting $H = 20$ km and $\beta = 0.084$. In the central region of the umbra the computed density is equal to the measured value and it is nearly neutral as it was during this eclipse in December 1963.

In a recent paper MATSUSHIMA [42] analysed in the same way two other eclipses, that in June 1964, measured by BOUŠKA and MAYER [43] and the eclipse in December 1964 measured by MATSUSHIMA. The values of β for the three above-mentioned eclipses were then compared with the astronomical determinations of visual extinction by MORENO and STOCK [44] at Cerro Tololo (2.2 km) and by PRZYBYLSKY [45] at Mt. Bingar (0.5 km). The results (Fig. 1.3.23) are very satisfactory showing a good agreement of both methods.

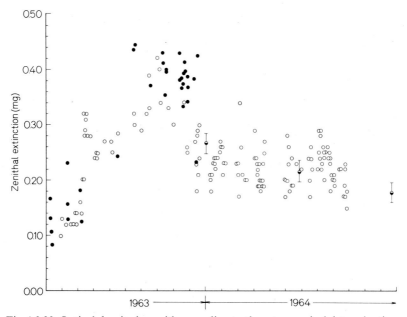

Fig. 1.3.23. Optical density in zenith according to the astronomical determinations compared with eclipse values (3×)

1.3.12. Global Intensity of the Eclipsed Moon

Measurements of the global intensity or luminosity of the Moon are very easy to realize by different photometrical methods from the visual comparison of reflected images on metallized glass spheres [46] to more elaborate photoelectrical measurements [47].

Unfortunately, their interpretation is not promising due to the fact that the global luminosity is expressed by the integral

$$L = k \int ae\,dq \qquad (1.135)$$

where a is the albedo of a lunar element dq and e

$$e = 10^{-D} \qquad (1.136)$$

is its illumination. Irregular repartition of the albedo over the lunar disk makes any simple interpretation of the above integral in terms of the shadow density D very difficult.

Some examples of the global light curves are given in Fig. 1.3.24. The lack of symmetry may be caused — if the extinction has been correctly accounted for — by two factors:

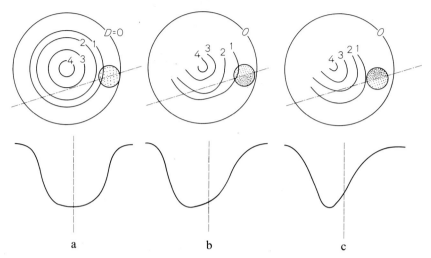

Fig. 1.3.24. Global intensity of the eclipsed Moon. Above: The situation of the Moon in the shadow. Below: The corresponding global intensity curves (situations b and c) are ambiguous in respect to the isophote forms

i) *Unequal repartition of the albedo* i.e. that of seas and continents during both halves of the eclipse.

ii) *Asymmetry of the density curve* as found during some eclipses. According to the circumstances, both the above factors may work in the same or in the contrary sense.

The global luminosity curve may supplement in some cases the density curve signalizing by its asymmetry the asymmetry of the density curve but only in the case, when the albedo repartition works in the opposite sense to the observed asymmetry of the global curve.

1.3.13. Surveyor III Eclipse Observation from the Moon [48]

During the Surveyor III mission to the Moon the lunar eclipse 1967 IV 24 took place and two series of pictures (20) were obtained at approximately 11 h 24 min and 12 h 01 min GMT. Surveyor landed at 2.3° S and 23.3° W in the SE region of the Oceanus Procellarum about 370 km south of Copernicus.

The survey television system had a camera mounted vertically and fed by a mirror which can be adjusted in azimuth and altitude. Three

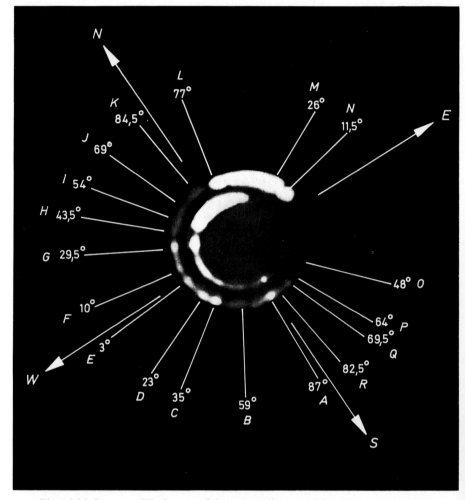

Fig. 1.3.25. Surveyor III pictures of the solar eclipse April 24, 1967 observed from the Moon. The first image taken at 11 h 24 min is reduced in size and superimposed upon the second image taken at 12 h 01 min

filters: blue with maximum transmission at 4500 Å, green 5500 Å and red 5900 Å were used.

The pictures obtained (Fig. 1.3.25) show a very bright refraction image of the Sun. They are saturated and consequently stretched out of the correct width, which would be only a small fraction of the terrestrial radius. In addition, a scattering halo of "beaded" appearance is visible round the Earth's limb. Comparing both series of images separated by

37 min in time, we find the shift of the refraction image due to the motion of the Moon across the shadow and almost unchanged positions of the light "beads" in the scattering halo, due probably to the gaps between the clouds along the shadow terminator. This conclusion is confirmed partly by the analysis of the cloudiness observed from the satellite ESSA-3 but, unfortunately, only on the day preceding the eclipse so that some shifts of cloud systems cannot be excluded.

Even when the intensity ratio between the saturated refraction image and the normally exposed scattering image cannot be obtained with any reliability, the importance of the latter image seems to us to be a little greater as it could be predicted from the theory. Therefore, a correct photometry of the phenomenon is badly needed before we can reach any definitive conclusion.

Bibliography

1. DANJON, A.: Ann. Obs. Strasbourg **2**, 1 (1928).
2. GUTH, VL., and F. LINK: J. Obs. **19**, 129 (1936).
3. WALKER, M.F., and G. REAVES: Publs. Astron. Soc. Pacific **69**, 153 (1957).
4. CIMINO, M., and A. FRESA: R.C. Accad. Lincei **25**, 58 (1958).
5. SANDULEAK, N., and J. STOCK: Publs. Astron. Soc. Pacific **77**, 237 (1965).
6. LINK, F.: Catalogue, Publs. Astron. Inst. Prague **29** (1956).
7. DANJON, A.: Compt. rend. **173**, 706 (1921).
8. CABANNES, J., et J. DUFAY: J. phys. (Paris) **8**, 125 (1927).
9. LINK, F.: Compt. rend. **196**, 251 (1933); — Ann. Aph. **9**, 227 (1946).
10. DOBSON, H.M.B.: Gerlands Beitr. Geophys. **24**, 13 (1929).
11. PAETZOLD, H.K.: Z. Naturforsch. **5**a, 661 (1950); **6**a, 339 (1951); **7**a, 325 (1952).
12. VIGROUX, E.: Ann. Aph. **17**, 399 (1954).
13. LINK, F.: Bull Obs. Lyon, **11**, 229 (1929).
14. HAUSDORFF, F.: Ber. Verh. Sächs. Akad. Wiss. **47**, 401 (1895).
15. BAUER, E., et DANJON A.: Compt. rend. **176**, 761 (1923).
16. ABBOT, C.G.: Ann. Astrophys. Observ. Smiths. Inst. **1**—**4** (1900—1922).
17. LINKE, F.: Handbuch der Geophys., Bd. 8/6, S. 256 1943.
18. GÖTZ, W.W.P.: Met. Z. **57**, 415 (1940).
19. LINK, F.: Gerlands Beitr. Geophys. **60**, 139 (1943).
20. LINK, F.: Ann. geophys. **4**, 225 (1948).
21. VASSY, E., et F. RÖSSLER: Compt. rend. **254**, 2041 (1962).
22. RÖSSLER, F.: Space Research **8**, 636 (1968).
23. JUNGE, C.E.: J. Met. **18**, 81 (1961).
24. LINK, F.: BAC **2**, 1 (1950).
25. MILLIKAN, R.A.: Phys. Rev. **21**, 217 (1923).
26. DAVIES, C.N.: Proc. Phys. Soc. (London) **57**, No. 322 (1945).
27. MURGATROYD, R.J., and F. SINGLETON: Quart. J. Met. Soc. **87**, 125 (1962).
28. GREENSTEIN, J.L.: Harv. Circ. No. 422 (1937).
29. ŠVESTKA, Z.: BAC **2**, 41 (1950).
30. BOWEN, E.G.: Australian J. Phys. **6**, 480 (1953).
31. GUTH, VL., u. F. LINK: Z. Aph. **18**, 207 (1939).
32. — — Z. Aph. **20**, 1 (1941).

33. LINK, F.: Stud. Geod. Geoph. **5**, 64 (1961).
34. DUFOUR, CH.: L'Astronomie **13**, 115 (1889).
35. FLAMMARION, C.: L'Astronomie **3**, 401 (1884).
36. SYMONS, G.L.: Report of Krakatoa Committee, London 1888.
37. HANSEN, J.E., and S. MATSUSHIMA: J. Geophys. Research **71**, 1073 (1966).
38. MEINEL, M.P., and A.B. MEINEL: Science **142**, 582 (1963).
39. HOGG, A.R.: Australian J. Sci. **26**, 119 (1963).
40. MOSSOP, S.C.: Nature **203**, 824 (1964).
41. GREEN, A.E.S.: Appl. Opt. **3**, 203 (1964).
42. MATSUSHIMA, S., J.R. ZINK, and J.E. HANSEN: Astr. J. **71**, 103 (1966).
43. BOUŠKA, J., and P. MAYER: BAC **16**, 252 (1965).
44. MORENO, H., and J. STOCK: Publs. Astron. Soc. Pacific **76**, 55 (1964).
45. PRZYBYLSKI, A.: Acta Astr. **14**, No. 4 (1964).
46. RICHTER, N.: Z. Aph. **21**, 249 (1942).
47. ANDRUSZEWSKI, S.: Acta Astr. **2**, 53 (1934).
48. Surveyor III-A Preliminary Report, NASA SP-146 (1967) 50.
49. EPSTEIN, P.S.: Phys. Rev. **23**, 710 (1924).

1.4. Lunar Luminescence
1.4.1. Simplified Theory of the Penumbra [1]

Photometric theory of the penumbra is reduced in the first approximation to the theory of a partial eclipse of the Sun by the Earth. It seems, therefore, to be a trivial phenomenon and its theory may be regarded as purely descriptive without any chance of throwing a new light on the aeronomy or lunar astronomy. However, we shall endeavour to correct this superficial impression.

In the simplified theory we shall take advantage of the classical theory of eclipsing variable stars. At first we define the ratio of angular radii of the Sun and of the Earth viewed from the Moon

$$R'_\odot/\pi_{\mathbb{C}} = k. \quad (1.137)$$

Its limits are $0,25 < k < 0,30$ and we may adopt the mean value $k = 0.28$. Then we consider the fractional loss of light α connected with the optical density by the relation

$$D = -\log(1-\alpha). \quad (1.138)$$

In the theory of eclipsing variables we consider the function named as the "geometrical depth of the eclipse" related to the angular distance of both bodies by

$$\gamma' = \pi_{\mathbb{C}} [1 + k\, p(k, \alpha, \kappa)]. \quad (1.139)$$

The depth p has been extensively tabulated by TSESEVICH [2]. We can finally replace the angular distance γ' by the distance from the edge of the geometrical shadow to obtain the formula

$$\gamma'' = \pi_\odot \{109.05[1+p(k, \alpha, \kappa)] - 1\} \quad (1.140)$$

which is nearly independent of the lunar parallax. From this formula and with the aid of Tsesewich's tables we obtain the following table of the density of the penumbra for different degrees of darkening κ.

Table 1.20. *Distances from the edge of the geometrical shadow*

D	$\kappa=0.0$ (')	0.2 (')	0.4 (')	0.6 (')	0.8 (')	1.0 (')
0.05	26.1	25.9	25.7	25.5	25.2	24.9
0.10	20.6	20.5	20.4	20.2	20.1	19.9
0.30	15.1	15.1	15.2	15.2	15.2	15.3
0.40	12.6	12.7	12.8	12.9	13.0	13.1
0.52	10.0	10.2	10.3	10.5	10.7	10.9
0.70	7.4	7.6	7.8	8.0	8.3	8.5
0.82	6.0	6.1	6.4	6.6	6.9	7.2
1.00	4.4	4.6	4.8	5.1	5.4	5.7
1.10	3.8	4.0	4.1	4.4	4.7	5.1
1.22	3.1	3.2	3.4	3.6	3.9	4.3
1.40	2.3	2.4	2.6	2.8	3.1	3.4
1.52	1.8	2.0	2.1	2.3	2.6	3.0

1.4.2. Complete Theory of the Penumbra

The above simplified theory neglects the presence of the atmosphere. Its influence is twofold. The atmosphere slightly increases the actual value of $\pi_{\mathcal{C}}$ and consequently also the density of the penumbra. Then the density of the shadow can be diminished by the formation of the refraction image of the Sun. It is, however, very difficult to evaluate the exact balance between these opposite effects and only a complete theory can decide it [3].

In elaborating this theory, we must take into account the actual value of limb darkening at the extreme part of the solar radius which differs slightly from the assumed formula [Eq. (1.16)]. For the limb darkening at $R/R_\odot$ 0.99 we used the values obtained by HEYDEN, as already stated (1.2.3). The structure of the upper atmosphere (O_3, high absorbing layer) also enters fully into the final results. Respecting these factors we have proceed with the integration [Eq. (1.12)] as in the umbra, the only difference being that in the outer part of the integration interval we put $T(h_0) = 1$.

The form of the illumination integral is given in Fig. 1.4.1. The curves show the above-mentioned influences of the solar limb and the structure of the high atmospheric layers. The resulting densities of the penumbra are given in the following table.

From this table it is evident that the atmosphere increases the density of the penumbra as compared with the simplified theory.

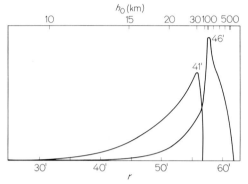

Fig. 1.4.1. Integrals of the illumination in the penumbra, the 41' curve is multiplied by 50

Table 1.21. *Mean densities of the penumbra*

γ''	O	C	$O-C$	E_0	E_c	E_c-E_0
0' 00''	2.87	2.74	+0.13	0.0013	0.0017	−0.0004
20''	2.64	2.61	+0.03	0.0023	0.0025	−0.0002
40''	2.35	2.46	−0.11	0.0045	0.0035	+0.0010
1' 00''	2.10	2.31	−2.21	0.0079	0.0049	+0.0030
20''	1.90	2.15	−0.25	0.0126	0.0071	+0.0055
40''	1.73	2.00	−0.27	0.0186	0.0100	+0.0086
2' 00''	1.60	1.85	−0.25	0.0251	0.0141	+0.0110
20''	1.52	1.70	−0.18	0.0302	0.0200	+0.0102
40''	1.44	1.56	−0.12	0.0363	0.0275	+0.0088
3' 00''	1.34	1.44	−0.10	0.0457	0.0363	+0.0094
20''	1.24	1.35	−0.11	0.0575	0.0447	+0.0129
40''	1.19	1.28	−0.09	0.0646	0.0525	+0.0121
4' 00''	1.10	1.21	−0.11	0.0794	0.0617	+0.0178
20''	1.04	1.14	−0.10	0.0912	0.0724	+0.0198
40''	0.99	1.10	−0.11	0.1023	0.0794	+0.0229
5' 00''	0.96	1.05	−0.09	0.1096	0.0891	+0.0205
6' 00''	0.85	0.92	−0.07	0.1413	0.1202	+0.0211
7' 00''	0.73	0.81	−0.08	0.1820	0.1549	+0.0271
8' 00''	0.66	0.72	−0.06	0.2188	0.1905	+0.0283
10' 00''	0.54	0.56	−0.02	0.2884	0.2754	+0.0130
15' 00''	0.31	0.31	0.00	0.4898	0.4898	0.0000
20' 00''	0.16	0.12	+0.04	0.6918	0.7586	−0.0668
25' 00''	0.07	0.05	+0.02	0.8511	0.8913	−0.0402

1.4.3. Comparison with Observations

The first comparison of observations with simplified theory was made by DANJON [4] for the eclipse October 16, 1921. He found a little light excess in the inner part of the penumbra and attributed it to the refraction image of the Sun. However, this interpretation is not suitable and hence we must look for another cause.

It was necessary at first to confirm this light excess from a number of eclipses. This was done [5] for 10 eclipses (1921 – 1943) and completed later [3] with a total of 18 eclipses (1921 – 1957). There does indeed exist a light excess in the inner part and also a little light defect in the outer part of the penumbra. The median values of the observed penumbral density are given in the Table 1.21 and represented in the Fig. 1.4.2. The maximum of excess 0.27 in density or nearly $2 \times$ in the illumination occurs at 1.7' from the edge of the geometrical shadow.

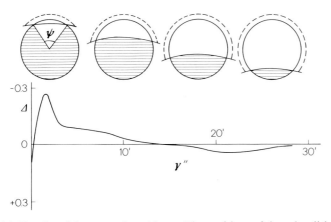

Fig. 1.4.2. Density of the penumbra. Above: The positions of the solar disk behind the Earth. Below: The $O-C$ curve; the differences between the observed and computed density of the penumbra

1.4.4. Interpretation of the Light Excess

Two trivial explanations of this rather unexpected phenomenon have been proposed. The first is in the scattered light from the Earth's atmosphere. To discuss it we shall assume the most favourable conditions:

a) The Earth's atmosphere appears as bright as a brilliant day sky veiled by cirri whose luminance amounts to 10^4 cd/m².

b) The height of this supraluminous atmosphere is of 120 km.

Both these assumptions are greatly overestimated.

The illumination produced by this atmospheric fringe around the dark disk of the Earth will be

$$\eta = \text{luminance} \times \text{solid angle} = 10^4 \times 3 \times 10^{-5} = 0.3 \text{ lux} \quad (1.141)$$

The actual illumination at the edge of the umbra ($D=3$) is about 100 lux or somewhat $300 \times$ larger than the parasitic maximum light scattered by the atmosphere.

The second trivial explanation is based on the opposition effect.

It is known that the luminance of the lunar surface increases very rapidly towards the opposition (Fig. 1.4.3). This so called opposition

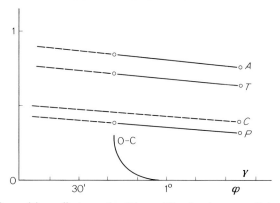

Fig. 1.4.3. Opposition effect on the Moon. The luminances of ARISTARCH (A), TYCHO (T), COPERNICUS (C) and PLATO (P) as a function of the phase angle φ [6]; $O-C$ the mean curve of densities in the penumbra

effect is attributed to the particular structure of lunar soil with many little holes casting the shadow outside the opposition [6]. It was, therefore, objected by BARBIER [7] and repeated by others that the curve of light excess found in the penumbra is simply due to this effect while the true opposition can occur only during the lunar eclipse.

In reality the resemblance of both curves i.e. of the light excess and the opposition effect is only superficial. This resemblance disappears when we trace both curves in the same scale (Fig. 1.4.3). The rise of the light excess is more rapid than that of the opposition effect. We must also seek for another explanation.

1.4.5. Lunar Luminescence

After the discovery of the light excess in the penumbra and with the difficulty of explaining it by the above minor effects, we have adopted as the natural explanation [1] the lunar luminescence whose existence has been confirmed later by other methods.

Lunar Luminescence

The existence of the lunar luminescence is a very plausible phenomenon as the great majority of known minerals show it under excitation by corpuscular or shortwave radiations. This excitation exists on the Moon in the absence of any perceptible atmosphere and has the same intensity as at the limits of our atmosphere. We may, therefore, expect some luminescence of the lunar surface whose low albedo in the visible spectrum favours the luminescent component in comparison with the diffused solar light.

The separation of these two components is possible by several methods. The first presents itself in the penumbra. If we consider [1] the shortwave electromagnetic radiation responsible for the excitation then its source should mainly be concentrated on the solar limb and in the low corona. As it results from simple geometrical consideration (Fig. 1.4.2) for a given point of inner penumbra the relative occulted part of the photosphere e is greater than that of the solar limb given by

$$\varepsilon = \frac{\psi}{360°}. \qquad (1.142)$$

The luminance of a lunar plage in the penumbra at the distance γ'' from the edge of the shadow will be

$$b = (1-l)\,e + \varepsilon\,l = e + l(\varepsilon - e) \qquad (1.143)$$

where l is the fraction of the luminescent component in the total light ($=1$). The fraction e is given in the table and we have the following comparison.

It may be noted that the difference $O-C$ is given by $l(\varepsilon-e)$ and from it the luminescent component may be estimated. In this way we get in the inner penumbra $l=5-10\%$, i.e. there is some $5-10\%$ of the luminescent component on the light of the full Moon outside the eclipse.

In the outer penumbra beginning approximately at $\gamma'' > 16'$ there should be a defect of light in accordance with the negative sign of $O-C$ in the Table 1.21. The observed differences $O-C$ seem to follow this inversion (Fig. 1.4.2).

Table 1.22

γ''	2.5′	5′	10′	15′	20′	25′	30′
ε	0.21	0.30	0.42	0.53	0.62	0.76	0.91
e	0.03	0.10	0.28	0.49	0.69	0.87	0.98
$\varepsilon - e$	+0.18	+0.20	+0.14	+0.04	−0.07	−0.09	−0.07

If the corpuscular radiations are the exciting agent of the lunar luminescence, we have some reason to believe that their trajectories are different from the light rays and the corpuscular occultation of the solar disk in the penumbra can be smaller than the light occultation. However, it is as yet very difficult to calculate the form of these trajectories but the above discussion leads us to the assumption that at least a part of exciting radiation comes from the solar limb.

1.4.6. Fluctuations of the Global Luminosity of the Moon

The second method for the separation of the luminescent component gives us the large fluctuation of the exciting solar radiation and the resulting fluctuation of the global luminosity of the Moon. Both short-wave and corpuscular radiation are known to be largely variable with solar activity. We may, therefore, expect to find the corresponding fluctuation in the lunar luminescence even though largely diluted in the diffused light. The first attempt to detect these variations was made using the valuable series of lunar measurements by ROUGIER [8]. He has obtained a mean curve of lunar luminosity as the function of the phase angle. This curve was traced from the individual measurements obtained by exact photoelectric photometry of Bouguer's straight lines and their extrapolations outside the atmosphere. The deviations of individual points from the mean curve can also give the transient fluctuations of the global lunar light [9].

These deviations Δm in stellar magnitudes were confronted with simultaneous variations of solar constant ΔC obtained by ABBOT [10] and a significant correlation (Fig. 1.4.4) of both quantities

$$r = -0.438 \pm 0.082 \text{ (m.e.)}; \quad n = 94 \qquad (1.144)$$

has been found.

The negative sign of the correlation coefficient is simply due to the definition of stellar magnitudes and from a physical view point must be taken as positive. The slope of the regression lines (Fig. 1.4.4) indicates further that the variation of 1% in the global solar radiation, i.e. in the solar constant, has in consequence a fluctuation of about 26% in the global light of the Moon. This is, after all, an analogy of a well-known fact. ABBOT [11] found that the ratio of spectral variation to the solar constant variation increases towards shorter wave lengths; at the limit 3500 Å of the accessible spectrum its value attains 6. A further increase at 3100 Å seems to exist according to PETTIT [12]. Finally the ionospheric measurements influenced by very short solar radiations show still larger fluctuations with solar activity.

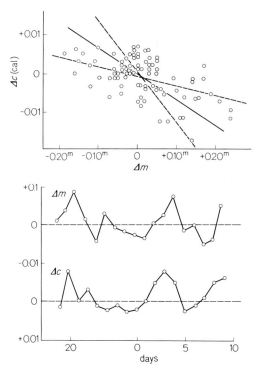

Fig. 1.4.4. Correlation between the variation of solar constant Δc and the deviations Δm of the Moon's magnitude from the mean curve. Above: The normal correlation graph. Below: The 27 day recurrence

Let us quote some similar conclusions by GEHRELS et al. [13] obtained by photoelectric photometry and polarimetry of the lunar surface:

i) The average lunar surface brightness at high solar activity 1958/59 was greater than at low solar activity 1963/64 by a factor of 1.14 ± 0.014.

ii) From polarimetric measurements this factor would be 1.12 ± 0.02.

iii) Some individual measurements show for instance the increase of 36% (November 17, 1956) or 27% (November 18, 1956) above the average level 1963/64.

iv) Lyot's polarimetric results [14] show with respect to February 29, 1924 the variation of the surface brightness $+16\%$ on November 4, 1923 and -10% on April 27, 1924.

The doubts on the light excess in the penumbra as the consequence of the lunar luminescence expressed by GEHRELS in connection with opposition effect have already been rejected (1.4.4).

1.4.7. Recent Work on Lunar Luminescence

The light excess in the penumbra was confirmed by numerous Italian measurements made by CIMINO, FORTINI, GIANNUZI and FREZA [39]. Unfortunately this method is limited only to some rare moments of eclipses. Therefore, we proposed as early in 1948 an independent method working outside eclipses, namely, the method of the central intensity of Fraunhofer lines in the lunar spectrum [15].

We measure the ratio $R = I_0/I_1$ of the central intensity to that of the adjacent continuum in lunar and solar spectrum with the same instrument and if possible under strictly identical conditions. Due to the luminescence band superimposed over the Fraunhofer line in lunar spectrum its profile will be shallower in comparison with the solar line (Fig. 1.4.5). The rate of the luminescence is then defined by

$$\rho = \frac{R_{\mathbb{C}} - R_{\odot}}{1 - R_{\mathbb{C}}}. \tag{1.145}$$

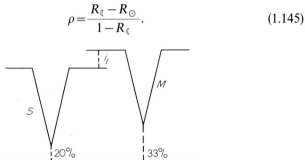

Fig. 1.4.5. Central intensity of the Fraunhofer line in solar (S) and lunar (M) spectrum, i_l the intensity of the luminescence

The method of the central intensity of Fraunhofer lines is of wide use in respect of time, wavelength and location on the Moon as has been proved by numerous results obtained up to the present time [16–20]. The ordinary spectral resolution was recently substituted by BLAMONT and CHANIN [21] by magnetic sweeping using the optical resonance according to BLAMONT [22]. The results obtained on Mare Tranquilitatis and Tycho in red 7699 Å show a curious decrease of ρ in progress of lunar day which could be related with the temperature. A similar phenomenon has been observed in the laboratory by SUN and GONZALES [23].

Finally the well known phenomenon of the emission in the crater Alphonsus discovered by KOZYREV [24] may be, according to ÖPIK, not an erruption of volcanic gas but simply the transient luminescence of the central peak. He wrote this about it [25]: "The reported emission ... is strictly confined to the illuminated portion of the peak, of a sharply bounded width of about 4 km. With molecular velocity of 0.5 km/sec during the 30 min of exposure the gases of the 'eruption' would have spread over a radius of 900 km. An interpretation... could be that of fluorescence of the sunlit peak."

1.4.8. Brightness of Lunar Eclipses and Solar Activity

The foregoing discussion gives us a better understanding of the variations in brightness of lunar eclipses disclosed by DANJON [26, 27] in relation to the solar activity. At that time (1922) good measurements of the shadow density were practically non-existent and only relatively numerous verbal descriptions of lunar eclipses were known. In dealing with this qualitative material DANJON established a relative scale of luminosity as follows.

Table 1.23. *Danjon's luminosity scale*

Degree	Description
0	Very dark eclipse. Moon hardly visible, especially at the mid totality
1	Dark eclipse, grey to brown colouring, details on the disk hardly discernible
2	Dark red or rust-colored eclipse with dark area in the center of the shadow, the edges brighter
3	Brick-red eclipse, the shadow often bordered with a brighter yellow zone
4	Orange or copper-colored very bright eclipse with bluish bright edge

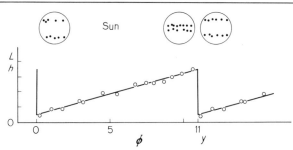

Fig. 1.4.6. Variations of the luminosity of eclipses with solar activity. Above: The positions of active zones on the solar disk

Both the luminosity and the color enter in the above definitions.

The first application of this scale to 70 eclipses between 1823–1920 showed an interesting relation of the luminosity L with the solar activity during the 11-year cycle (Fig. 1.4.6). This was completed by DE VAUCOULEURS [28] with other 47 eclipses between 1893–1943 with the same results and also the recent eclipses are in agreement with Danjon's saw-toothed curve. At the solar minimum ($\Phi=0$) the curve shows a pronounced discontinuity. The eclipse is brightest shortly before the minimum, i.e. during the old solar cycle, and briefly after the minimum with the new solar cycle, the luminosity drops suddenly to a low level. Between the two minima the luminosity varies nearly linearly.

Danjon's work encountered many objections. MAUNDER's objection [29] relating to the incorrect use of partial eclipses was not justified

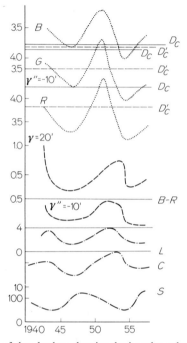

Fig. 1.4.7. Variations of the shadow density during the solar cycle for the blue (B), green (G) and red (R) color at $\gamma'' = -10'$. D_C is the computed density in Rayleigh atmosphere, D'_C the same with O_3

as DANJON included in the statistics only the bright partial eclipses and not the dark ones which can be dark simply by contrast with the bright lunar crescent.

Fisher's objections were raised on the basis of new independent statistics. FISHER established a different luminosity scale based on simple visibility of lunar details in telescopes of increasing aperture and investigated the eclipses in the period 1860–1922. In this manner [30]. FISHER confirmed neither the 11-year relation nor the dependence of the luminosity of total eclipses on the parallax. This last relation based on the theory (1.2.12) was nevertheless confirmed by observations [31] expressed, however, in Danjon's scale. It seems to us, therefore, that

Fisher's scale, totally neglecting the role of the magnification used by the observers, is less suitable for eclipse purposes than Danjon's scale.

At the present time we have the possibility to verify Danjon's relation by photometric measurements of the shadow density. Eighteen eclipses (1921–1957) have been used and the results on the whole confirm Danjon's relation [32]. The density in the blue, green and red is represented in Fig. 1.4.7 as the function of time. Furthermore the color of the umbra given by the differences of blue-red density, the luminosity L in Danjon's scale, the brightness C of the earth-light on the Moon and sunspots numbers S are represented. The relation of the shadow density to solar activity is clearly seen and the color drops abruptly after the solar minimum in 1954.

A part of the density curves placed on the descending branch of the solar activity curve lies over the theoretical limit of Rayleigh atmosphere with the ozone or even without it. There is consequently a supplementary illumination as is shown for the very bright eclipse of September 26, 1950 in the following table.

Table 1.24. *Eclipse 1950 IX, 26*

Density in Rayleigh atmosphere	4.60	3.56	3.02
Density due to O_3	0.03	0.33	0.43
Computed density	4.63	3.89	3.45
Observed density	3.68	3.40	3.09
$O-C$	−0.95	−0.49	−0.36
Observed illumination $\times 10^6$	209	398	813
Computed illumination $\times 10^6$	23	129	355
$(O-C) \times 10^6$	+186	+269	+458

The actual solar activity during each eclipse is only loosely connected with the phase within the 11-year cycle. Therefore, MATSUSHIMA [33] took the geomagnetic planetary index K_p as an argument for his researches on the basis of the same eclipse material [32]. He found (Fig. 1.4.8) a monotonic decrease of the shadow density with increasing K_p and an analogous correlation was found for the color given by the differences Blue-Red or Blue-Green, the eclipsed Moon being redder when K_p was high.

Danjon's relation may be explained [1] on the basis of the lunar luminescence excited by solar corpuscular radiation. The streams of solar corpuscles emitted radially have a greater probability of reaching the Moon when their sources are in low heliographic latitudes. According to Spörer's law, their latitude is high just at the beginning of the solar cycle and sinks progressively to the equator till the next solar minimum. In the turning-point to the next cycle the active centers jump abruptly to high latitudes (Fig. 1.4.6). The lunar luminescence excited by solar corpuscules, and consequently, the luminosity of lunar eclipses follows this course and the discontinuity on Danjon's curve is the image of the jump on the latitude curve illustrating Spörer's law. The relation of the luminosity to the planetary geomagnetic index K_p is obvious as it is at same time the index for the intensity of corpuscular streams in the terrestrial space.

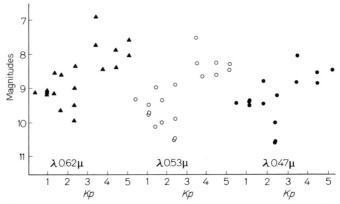

Fig. 1.4.8. Correlation of the shadow density (in magnitudes) at $\gamma'' = -10'$ with the geomagnetic planetary index K_p

1.4.9. Danjon's Relation and the Solar Cycle

Danjon's relation gives us the opportunity to determine the dates of solar minima in the past with the aid of the lunar eclipse, as it was actually exploited by DANJON himself [27]. He took all records of luminosity drops of lunar eclipses for solar minima, which are given under his name in the following table.

The mathematical representation of these data leads to the following expression

$$\text{Min} = 1584.8 + 10.87\, E + 1.7 \sin 2\pi \frac{t - 1608}{136} \tag{1.146}$$

$$E = 0, 1, 2, 3, \ldots.$$

Before 1823 there are big discrepancies between the eclipses and solar minima according to WOLF. But we must agree with DANJON that the sunspots before that year have been obtained, not from routine observations inaugurated by SCHWABE in 1823, but, from a compilation

Table 1.25. *Solar minima*

D	E	W	D − W	D − Eq. (1.146)
1583	0			−0.2
1606	2	1610.8	−4.8	−0.4
1618	3	1619.0	−1.0	−0.2
1672	7	1666.0	+6.0	+2.5
1703	11	1698.0	+5.0	+0.2
1713	12	1712.0	+1.0	−0.4
1725	13	1723.5	+1.5	+0.6
1737	14	1734.0	+3.0	+0.8
1749	15	1745.0	+4.0	+0.2
1760	16	1755.2	+4.3	+0.2
1782	18	1784.7	−2.7	−1.1
1812	21	1810.6	−1.4	−0.1
1823	22	1823.3	−0.3	+0.6
1834	23	1833.9	+0.1	0.0
1844	24	1843.5	+0.5	+1.2
1856	25	1856.0	0.0	+1.1
1868	26	1867.2	+0.8	+1.5
1878	27	1878.9	−0.9	−0.2
1889	28	1889.6	−0.6	−0.8
1901	29	1901.7	−0.7	−0.4
1912	30	1913.6	−1.6	−0.6
1923	31	1923.6	−0.6	−0.3

Solar minima: D according to Danjon's eclipse observations, W according to Wolf's compilations, E serial number.

of fragmentary observations collected by WOLF from old records. The discovery of sunspots in 1611 would probably not have happened if it had been the year of solar minimum as indicated by WOLF. Therefore, we prefer before 1823 the eclipse data to Wolf's compilations. After 1823 the agreement of both series becomes suddenly very good, especially when we add, according to BELL and WOLBACH [34], 0.4 year to the first term of the Eq. (1.145). These authors extrapolated it for the future in order to obtain the next solar minima in 1974.9; 1985.7; 1997.0; 2008.6; and 2020.0.

One may be intrigued by the average length of the solar cycle of 10.87 years in contrast with the generally quoted value of 11.1 years. We must first of all point out that during the best observed interval from 1823 to 1954, where eclipse and solar minima closely agree, the average length of solar cycle is only 10.9 years. Secondly, independent researches on solar activity in the past on the basis of climatic variations [35] led to the average value of 10.85 years with the limits from 12.1 to 9.6 years. Danjon's relation merits, therefore, due attention in all tentative forecastings of solar activity.

1.4.10. Problematic Variations of the Penumbral Density

Besides the density variations in the inner part of the penumbra connected with the luminescence we shall discuss the conclusions by MARKOV and SCHEGOLEV [36] on penumbral variations at the distance $\gamma = 1.4\sigma$. The density at this distance might be greater by 0.8 at the solar minimum than during the rest of the 11-year cycle. Later, as the minimum eclipse on November 29, 1955 was clear beyond expectation, MARKOV

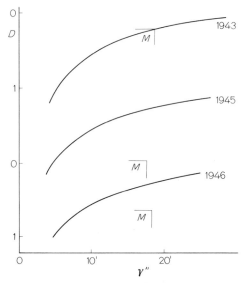

Fig. 1.4.9. Different penumbral curves according to ROUGIER compared with the values (M) adopted by MARKOV

modified the above conclusion stating that only the even solar minima are effective, and therefore, the assumed period of penumbral variations should be of 22 years [37].

Without waiting for any further modification of his hypothesis, we can show its weak observational and theoretical bases. The weak point of Markov's observational basis is the different sources used, so that the supposed variability may have its origin in systematic errors of the methods used for observations. To eliminate this source of errors, we [38] compared his results with three eclipses measured by ROUGIER and DUBOIS with the same cat's eye photometer (1.3.2). The curves of the penumbral densities are given in Fig. 1.4.9 where the values adopted by MARKOV and SCHEGOLEV are also marked. The first eclipse on August 15, 1943, according to MARKOV clear, agrees with ROUGIER. On the other hand, the two eclipses on December 18, 1945 and December 8, 1846, very dark according to MARKOV, are practically identical with Rougier's first eclipse. Markov's third dark eclipse on February 8, 1925 was unfortunately not measured by ROUGIER. Nevertheless, it seems difficult to accept the reality of such a great darkness in the penumbra.

MARKOV explaines the obscuration of the penumbra by the temporary pollution of the upper atmosphere produced by solar activity. If we represent the relative position of the Sun and the Earth as seen by the observer at the distance $\gamma = 1.4\sigma$ (Fig. 1.4.10), we realize that the total

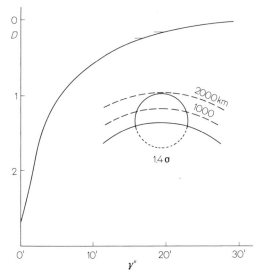

Fig. 1.4.10. Mean penumbral curve. The relative position of the Sun and the Moon is given at the distance $\gamma = 1.4\sigma$

opacity for grazing solar rays up to 1000 km or the partial transmission of 1/6 up to 2000 km are necessary to give the assumed variation of 0.8 in the penumbral density. That, however, seems as unrealistic for the height of the polluted layers as for the degree of the pollution.

Consequently, we can see that both theoretical and observational evidence conflicts with the possibility of any great variation of penumbral density at $\gamma = 1.4\sigma$, and it would be premature to accept the above hypothesis.

Bibliography

1. LINK, F.: Bull. Math. Phys. Tchécosl. **72**, 65 (1947).
2. TSESEVICH, V.: Bull. Inst. Astr. S.S.S.R. No. **50** (1940).
3. LINK, F.: BAC **9**, 169 (1958).
4. DANJON, A.: Comp. rend. **173**, 706 (1921).
5. LINK, F.: Comp. rend. **223**, 976 (1946).
6. DIGGELEN, J. VAN: Plan. Space Sci. **13**, 271 (1965).

7. BARBIER, D.: Photometry of lunar eclipses. In: Solar system III (ed. KUIPER) Chicago 1962, p. 249.
8. ROUGIER, G.: Ann. Obs. Strasbourg **2**, 3 (1933).
9. LINK, F.: Coll. Lyon, CNRS **1**, 308 (1947).
10. ABBOT, C.G.: Ann. Astrophys. Observ. Smiths. Inst. **6**, 85 (1942).
11. ABBOT, C.G.: Ann. Astrophys. Observ. Smiths. Inst. **6**, 164 (1942).
12. PETTIT, E.: Bull. Solar Phen. Zürich **24** (1933); **44** (1938).
13. GEHRELS, T., T. COFFEEN, and D. OWINGS: Astron. J. **69**, 850 (1964).
14. LYOT, B.: Ann. Obs. Meudon **8** (1929).
15. LINK, F.: Trans. Int. Astr. Un. Zürich 1948, **7**, 135 (1950); BAC **2**, 131 (1951).
16. DUBOIS, J.: Mem. Czech. Acad. Sci. **69** (1959); — J. phys. radium **18**, 13 (1957).
17. KOZYREV, N.A.: Izv. Crim. Astroph. Obs. **16**, 148 (1956).
18. GRAINGER, J.F.: Astr. Contr. Univ. Manchester **3**, No. 104 (1963).
19. SPINRAD, H.: Icarus **3**, 500 (1964).
20. SCARFE, C.D.: M.N. **130**, 19 (1965).
21. BLAMONT, J., et M.L. CHANIN: Compt. rend. **266**B, 514 (1968).
22. BLAMONT, J.: Compt. rend. **237**, 1320 (1963).
23. SUN, K.H., and J.L. GONZALEZ: Nature **212**, 24 (1966).
24. KOZYREV, N.A.: Physical observations of the lunar surface, Physics and Astronomy of the Moon (ed. KOPAL), p. 361. New York 1962.
25. OEPIK, E.J.: Irish Astr. J. **6**, 39 (1963).
26. DANJON, A.: Compt. rend. **171**, 127 (1920).
27. DANJON, A.: Compt. rend. **171**, 1207 (1920).
28. VAUCOULEURS, G.: Compt. rend. **218**, 655, 805 (1944).
29. MAUNDER, E.W.: J.B.A.A. **31**, 346 (1921).
30. FISHER, W.: Smiths. Misc. Coll. **76**, No. 9 (1924).
31. ŠVESTKA, Z.: BAC **1**, 48 (1948).
32. LINK, F.: BAC **11**, 13 (1960).
33. MATSUSHIMA, S.: Nature **211**, 1027 (1966).
34. BELL, B., and J.G. WOLBACH: Icarus **4**, 409 (1965).
35. LINK, F.: Plan. Space Sci **12**, 337 (1964).
36. MARKOV, A.V., i D.E. SCHEGOLEV: Izv. Pulkov. Obs. **151**, 34 (1953).
37. MARKOV, A.V.: Astr. Cirk. S.S.S.R. **306**, 4 (1964).
38. LINK, F.: BAC **17**, 161 (1966).
39. Acad. nazl. Lincei, Rend. **14**, 619 (1953); **17**, 211 (1954); **18**, 65, 173 (1955); **25**, 58 (1958).

1.5. Increase of the Shadow

1.5.1. Short History of the Shadow Increase

From the beginning of the 18th century astronomers realized that Hipparchos' formula [Eq. (1.1)] for the size of the umbra

$$\sigma = \pi_\odot + \pi_\mathrm{C} - R_\odot \tag{1.1}$$

needs a positive correction of some percents called the *shadow increase*. LAHIRE gave [1] its first value of $1/41 = 2.5\%$ according to observed times of contacts of lunar disk with the umbra. CASSINI [2] attributed this phenomenon to the terrestrial atmosphere and we can find a reference to it in the subsequent work of LEMONNIER [3], LALANDE [4] and LAMBERT [5].

Later came MAEDLER [6] with a mathematical method for determining the increase by the observation of transits of lunar craters through the edge of the umbra and SCHMIDT [7] continued in this way using numerous observations of his own. BROSINSKY [8] still using this method reduced 20 eclipses observed by different observers from 1776 to 1888. At the end of 19th century the improved lunar theory enabled HARTMANN [9] to introduce a perfected method and thus reduce 28 eclipses between 1802 and 1889. A half-century later KOSIK [10] established his own method based on the equatorial system of coordinates, while in the past the ecliptical system was used throughout. Kosik's method has been used in all subsequent work, as will be shown later.

1.5.2. Maedler's Method

The theory of lunar motions was not exact enough in the first half of the 19th century to calculate the precise position of the Moon for every eclipse. The main uncertainty concerned the longitude, the errors in the latitude and in the parallax being of minor importance. Therefore, MAEDLER worked out the method [6] best fitted to these circumstances. By observation we determine the duration of the eclipse for different craters and from it we derive the length of the chord circumscribed within the shadow (Fig. 1.5.1). Knowing further from the ephemeris the

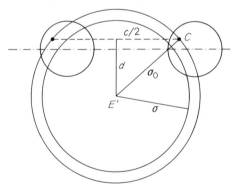

Fig. 1.5.1. Maedler's method for the determination of the shadow increase

minimal distance of the chord from the center of the umbra (=antisun) we compute the angular radius of the shadow by

$$\sigma_0 = \sqrt{d^2 + \left(\frac{c}{2}\right)^2} \qquad (1.147\text{a})$$

and finally the increase of the shadow

$$S = \frac{\sigma_0 - \sigma}{\sigma}. \qquad (1.147\,b)$$

The application of Maedler's method demands, of course, the complete observation of the eclipse, i.e. the entry and the exit of the craters a condition which cannot be always fulfilled. Furthermore, we are not able to distinguish between the increase of both sides of the umbra. The only merit mentioned above was, however, lost to the improvement of the lunar theory.

1.5.3. Hartmann's Method

HARTMANN considers separately the entry and the exit of each crater and computes for the corresponding moments the distance of the crater from the center of the shadow i.e. the actual radius of the shadow. The exact knowledge of lunar ephemeris is here necessary. Hartmann's formulae are developed in the ecliptical spherical coordinates currently used at that time [9].

1.5.4. Kosik's Method

KOSIK introduced the equatorial coordinates in the shadow increase investigations, namely the rectangular system of coordinates [10]. This makes the computations easy to verify and the method is also better adapted to the modern presentation of the lunar ephemeris.

KOSIK locates the origin of his rectangular coordinates system at the center of the Earth. The $+Oz$ axis is directed along the shadow axis towards its center, the $+Oy$ axis towards the north celestial pole and the $+Ox$ axis lies in the equatorial plane in the direction of lunar revolution around the Earth.

In this system the Moon has the following coordinates:

$$x_{\mathbb{C}} = \frac{\cos \delta_{\mathbb{C}} \sin (\alpha_{\mathbb{C}} - \alpha_1)}{\sin \pi_{\mathbb{C}}}$$

$$y_{\mathbb{C}} = \frac{\sin (\delta_{\mathbb{C}} - \delta_1)}{\sin \pi_{\mathbb{C}}} + 0.008726 (\alpha_{\mathbb{C}} - \alpha_1)^\circ \, x_{\mathbb{C}} \sin \delta_1 \qquad (1.148)$$

$$z_{\mathbb{C}} = \frac{\cos (\delta_{\mathbb{C}} - \delta_1)}{\sin \pi_{\mathbb{C}}} - 0.008726 (\alpha_{\mathbb{C}} - \alpha_1)^\circ \, x_{\mathbb{C}} \cos \delta_1$$

where $\alpha_{\mathbb{C}}, \delta_{\mathbb{C}}$ are the apparent right ascension and declination of the Moon, α_1, δ_1 analogous coordinates of the antisun and $\pi_{\mathbb{C}}$ the lunar parallax.

On the Moon each crater has its selenographical coordinates λ, β and $R_{\mathbb{C}}$ the distance from the center of the Moon. For the corresponding rectangular coordinates we have

$$x_0 = R_{\mathbb{C}} \cos \beta \sin \lambda$$
$$y_0 = R_{\mathbb{C}} \sin \beta \qquad (1.149)$$
$$z_0 = R_{\mathbb{C}} \cos \beta \cos \lambda.$$

These classical selenographic coordinates may be transformed into Kosik's system by means of the translation of the origin and the rotation of the axis according to the formulae

$$x = x_{\mathbb{C}} + a_x x_0 + b_x y_0 + c_x z_0$$
$$y = y_{\mathbb{C}} + a_y x_0 + b_y y_0 + c_y z_0 \qquad (1.150)$$

where the direction cosines of the axes Ox, Oy an Oz of the first system in the second system are given by

$$a_x = -\cos \lambda_\odot \cos P + \sin \lambda_\odot \sin P \sin \beta_\odot$$
$$b_x = \sin P \cos \beta_\odot$$
$$c_x = \sin \lambda_\odot \cos P - \cos \lambda_\odot \sin P \sin \beta_\odot$$
$$a_y = \cos \lambda_\odot \sin P - \sin \lambda_\odot \cos P \sin \beta_\odot \qquad (1.151)$$
$$b_y = \cos P \cos \beta_\odot$$
$$c_y = -\sin \lambda_\odot \sin P - \cos \lambda_\odot \cos P \sin \beta_\odot$$

where $\lambda_\odot, \beta_\odot$ are the selenegraphic coordinates of the Sun, and P the position angle of the lunar rotation axis projected into the tangent plane to celestial sphere at the eclipse place. All these quantities may be found in the modern ephemeris.

We prepare the reduction of an observed eclipse by computing for hourly intervals in U.T. the above mentioned quantities.

Then for each observed crater transit we interpolate them and compute the distance of the crater from the center of the shadow

$$\sigma_0 = \sqrt{x^2 + y^2} \qquad (1.152)$$

and the position angle from the equator

$$\log \psi = \frac{y}{|x|}. \qquad (1.153)$$

A small correction

$$\Delta \sigma_0 = 0.0046 (z - z_0) \qquad (1.154)$$

has to be introduced due to the fact that the crater generally lies nearer to the Earth than the center of the Moon.

1.5.5. Results on the Shadow Increase

Up to present time we have several determinations of the shadow increase:

1776 – 1888	20 eclipses	BROSINSKY, 1889 [8].
1802 – 1889	28 eclipses	HARTMANN, 1891 [9].
1889 – 1936	23 eclipses	LINK and LINKOVÁ, 1954 [11].

In addition some further eclipses have been reduced by KOSIK [10], KOEBCKE [12] and especially by BOUŠKA [13].

From a set of 57 eclipses in the last 150 years we have obtained the mean value of the shadow increase of 2.3% i.e. close to the conventional value of 2.0% adopted for the computation of the ephemeris.

Besides the increase of the shadow, KOSIK [10] disclosed in some well observed series the flattening of the shadow i.e. the dependence of the increase on the latitude of the effective terminator which is similar to that of the geoid but considerably larger. Here are the first determinations by KOSIK illustrating this fact:

Eclipse	Radius of the shadow	
1938 XI, 7	$43.67' - 0.353 \sin^2 P$	observed
	$42.95' - 0.183 \sin^2 P$	computed
	$0.72' - 0.170 \sin^2 P$	flattening
1939 V, 3	$42.90' - 0.456 \sin^2 P$	observed
	$42.12' - 0.183 \sin^2 P$	computed
	$0.78' - 0.273 \sin^2 P$	flattening

P is the position angle from the equator.

The flattening of the shadow manifests itself in photometrical measurements of the shadow at its edge as well as inside and outside this limit. There is a difference in the form of the density curves (Fig. 1.5.2) between the low and high latitudes, the former being steeper and their bend being located further away from the edge. In the penumbra there is a difference of the density in the same sense between low and high latitudes (Fig. 1.5.2) and its amount decreases with increasing distance from the shadow limit.

Our feeling is that further investigations in this field are badly needed either in the reduction of old observations or in the programming and practical execution of observations during future eclipses. Collaboration with amateur astronomers should be of great help.

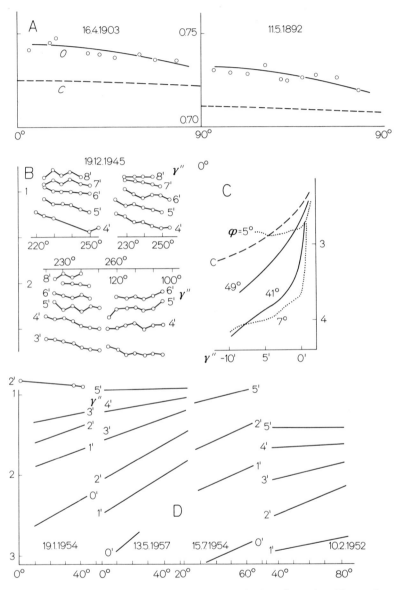

Fig. 1.5.2 A–D. Different manifestations of the shadow flattening. From above successively: A Shadow increase during two eclipses from the equator (0°) to the pole (90°), O observed and C computed values. B isophotes of the penumbra for different distances γ'' as a function of the position angle (90° or 270° = equator). C density curves at the periphery of the shadow at different latitudes, C is the computed curve for middle latitudes. D schematic variations of the density in the penumbra at different distances γ'' as a function of the latitude

1.5.6. Explanation of the Shadow Increase

The intriguing fact of the shadow increase known from the beginning of the 18th century led at the end of the 19th century HEPPERGER and SEELIGER to make a serious quantitative attempt to explain it theoretically. Both of them have independently computed the density curve of the shadow $D = f(\gamma)$ at its edge in order to examine if some particularity in the curve might be connected with the observed increase of about 50". Even though their efforts were handicapped by an insufficient knowledge of the atmospheric structure, nevertheless the principle used merits our attention. Their results are given in the Table 1.2.6.

Table 1.26. *Shadow densities*

γ (")	HEPPERGER		SEELIGER without		with limb darkening	
2460	2.424		2.653		2.820	
		38		45		34
2470	2.386		2.608		2.786	
		42		53		36
2480	2.344		2.555		2.750	
		49		67		40
2490	2.295		2.488		2.710	
		60		92		48
2500	2.235		2.396		2.662	
		83		121		61
2510	2.152		2.275		2.601	
		89		125		79
2520	2.070		2.150		2.522	
		92		114		91
2530	1.978		2.036		2.431	
		84		100		98
2540	1.894		1.936		2.333	
		80		89		96
2550	1.814		1.847		2.237	
				79		92
2560			1.768		2.145	

HEPPERGER [14] located the observed limit of the shadow at the inflexion point of the curve $D = f(\gamma)$ i.e. where $dD/d\gamma$ or the relative variation of the illumination $de/e : d\gamma$ attains its maximum. This principle agrees with the modern theory of the contrast elaborated by KÜHL [15]. The application of Hepperger's values (Table 1.26) leads to an increase of about 52", agreeing with observations. However, Hepperger's values were computed neglecting the limb darkening and within the atmosphere (1.2.22) which differs considerably from its actual structure. The agreement he found is, therefore, merely accidental.

SEELIGER [16] whose atmospheric model is more suitable (1.2.22) found with limb darkening and using the above criterion an increase of 70″, much greater than the observed value. This was probably the reason that led him to experiment with a rotating disk. He constructed a black-and-white painted disk (Fig. 1.5.3) whose central angle in the white part at each distance from the center was proportional to the illumination at the corresponding point of the shadow. When the disk was made to rotate rapidly, its luminance observed from a convenient distance appeared according to Talbot's law as bright as the luminance of the eclipsed Moon on the hazy border of the shadow. By means of a

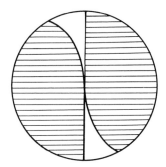

Fig. 1.5.3. Seeliger's disk for the simulation of the shadow

suitable micrometer different observers measured the diameter of the diffused disk and in such a manner SEELIGER derived an increase of about 50″, i.e. without any relation to the inflection point of the shadow curve. This finding led SEELIGER to the conclusion that the phenomenon has a purely physiological origin independent of the course of the density, at least if we accept the densities resulting from his theory.

Moreover, SEELIGER opposed the idea that the increase of the shadow might be the effect of the shadow cast by the atmosphere while the extinction, e.g. at 36 km, as he wrote, is nearly negligible and, therefore, the atmosphere should be there practically transparent. SEELIGER seems to have failed to realize the existence of the attenuation by the refraction and could not have anticipate the existence of any absorbing medium in the upper atmosphere.

1.5.7. Paetzold's Experiments

PAETZOLD has repeated Seeliger's experiments, improving both the theoretical basis and the arrangement of experiments [17]. As far as the latter is concerned the white screen receives the illumination from an electrical bulb so as to give the luminance of 0.01 stilb which we meet on the Moon at 3′ from the edge of the shadow. At a distance of 3 meters from the screen rotates the disk whose aperture is modelled according to the law $D=f(\gamma)$ adopted for the experiment. The observer looks from the distance of 20 meters towards the disk by means of a small telescope with $10\times$ magnification in order to simulate the observing conditions of the Moon with $30\times$ magnification. The limit of the shadow was measured with the aid of an ocular micrometer.

PAETZOLD adopted four different atmospheric models in order to compute the function $D=f(\gamma)$. Their characteristics together with the observed increase are given in the following Table 1.27.

Table 1.27

Model		Shadow increase (%)
I	Rayleigh atmosphere according to LINK [23] with ozone	2.44
II	Rayleigh atmosphere I without ozone	2.54
III	Rayleigh atmosphere as I up to 30 km, above 30 km the density gradient $\frac{1}{2}$ of its value + ozone	1.61
IV	Model as I with high absorbing layer at 90 km and zenithal density 0.012	3.19

On the other hand PAETZOLD computed also the values of $dD/d\gamma$ to see whether the location of its maximum corresponds to the observed increase of the shadow as is in fact the case.

As a result of such work one thing is clear, that the fluctuations of the observed increase cannot have their origin in the lower atmosphere. Notwithstanding certain differences of models I–III the resulting values of the increase show only slight variations. On the other hand, the presence of the high absorbing layer (model IV) modifies its value in a striking manner. The absolute value of the increase itself depends on the experimental conditions such as parasitic light or the intensity of the illumination without noting the color of the artificial shadow which differs from the natural conditions. Hence we cannot expect the agreement of both values but only the origin and location of the variations are significant.

1.5.8. Connection between the Shadow Increase and the Meteoric Activity

Paetzold's findings on the importance of the high absorption producing the fluctuations of the shadow increase seem to be confirmed by their connection with meteoric activity. ŠVESTKA and BOUŠKA [18] have shown the annual variation of the increase with the activity of meteoric showers. Their results, based on the old observational material reduced by HARTMANN [9], were greatly extended by the new reduction of eclipses [19] by Kosik's method. The complete curve embracing 57 eclipses 1804—1950 is shown in Fig. 1.5.4.

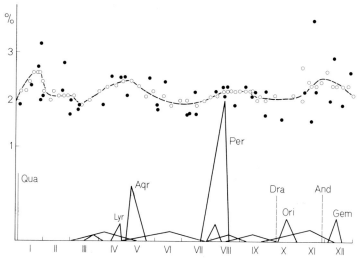

Fig. 1.5.4. Annual variation of the shadow increase compared with the activity of the principal showers [19]

The mean curve displays the connection with principal meteoric showers. However, we should be aware of the fact that, the shower activity is estimated according to the visible meteors while the shadow increase depends on the dust content of each shower which may be only loosely connected with the visual activity. According to recent estimations the ratio of invisible micrometeorites to visible meteors may be as great as $10^2:1$ depending further on the age of the shower [22].

The residence of meteoritic dust in the upper atmosphere is of short duration because of the rapid fall in these rarefied layers (1.3.8). A striking example is given by the eclipse on November 16, 1910 arriving just at the maximum of Leonids and exhibiting an unusually big increase of 3,7%.

1.5.9. Discussion of the Shadow Flattening

The amazing fact disclosed by KOSIK [10] and confirmed later by other investigations [11, 13] is the positive difference between the optical flattening of the shadow and the flattening of the geoid. On average the shadow is about 2—3 times more flattened than the Earth itself. Let us give some points of the discussion by SCHILLING [20]. It is clear that if the atmosphere is arranged in the form of concentric layers over the flattened Earth or if the atmosphere at higher levels follows the equipotential surfaces the difference between the Earth's and atmospheric flattening will be very small. Furthermore, satellite drag data have provided some indications of an equatorial bulge in the atmospheric density which is believed to be caused by the differential heating of equatorial and polar regions. However, this bulge is most apparent at 500 km and above, and becomes far smaller than the eclipse value under 200 km [21]. Thus SCHILLING concludes that the latitudinal variations of Rayleigh's atmosphere cannot account for the observed eclipse flattening.

SCHILLING adopts the explanation of the shadow increase by the presence of the high absorbing layer whose upper limit should be located at the average altitude $h_m = 80$ km. The immediate consequence is the difference $h_E - h_P$ of equatorial and polar height of the layer which depends on the adopted value of the flattening. In any case the difference of 25 or 41 km will be attained if we assume the flattenings 1/139 or 1/104. This difference might be also the flattening of the mesopause if its level is in some way connected with the upper limit of the high absorbing layer.

There are indeed some reasons for this hypothesis. The mesopause marked by the minimum of temperature is also the boundary between the underlying turbulent regime and the higher and more stable regime. Above the mesopause, due to the absence of convection, the fall velocity of dust particles will depend only on their size and the air density and according to our formulae [Eq. (1.119)] it should be relatively great. At the mesopause level with its temperature reversal and the corresponding discontinuity in density gradient the fall rate can be affected, as it can below the mesopause, where the turbulent regime can considerably check the fall. In other words there are great chances for the abrupt increase of dust concentration just at this level and lower. However, this rather qualitative explanation needs further quantitative examination.

Bibliography

1. LAHIRE, P.: Tabulae astronomicae. Paris 1707.
2. CASSINI, J.D.: Tables astronomiques, p. 34. Paris 1740.
3. LEMONNIER, O.C.: Institutions astronomiques, p. 251. Paris 1746.
4. LALANDE, J.F.: Astronomie **2**, 337. Paris 1782.
5. LAMBERT, J.: Briefwechsel. Berlin 1782.
6. MAEDLER, J.H.: AN **15**, 1 (1838).
7. SCHMIDT, J.F.I.: Der Mond, S. 34. Leipzig 1856.
8. BROSINSKY, A.: Über die Vergrößerung des Erdschattens. Berlin 1888.
9. HARTMANN, J.: Abhandl. sächs. Ges. Wiss., Math.-physik. Kl. **17**, 365 (1891).

10. Kosik, S. M.: Bull. Tashk. Obs. **2**, 79 (1940).
11. Link, F., u. Z. Linkova: Publ. Obs. Prague **25** (1954).
12. Koebcke, F.: Bull. soc. amis sci. Poznán **11**, 39 (1951); **13**, 385 (1954).
13. Bouška, J.: BAC **1**, 37, 75 (1948); **2**, 28 (1950); **7**, 85 (1956); **9**, 245 (1958); **11**, 145 (1960); **17**, 92 (1966).
14. Hepperger, J.: Sitzber. Akad. Wiss. Wien, Math.-physik. 54/II (1895).
15. Kühl, A.: Physik. Z. **29**, 1 (1928).
16. Seeliger, H.: Abhandl. bayer. Akad. Wiss. II. Kl. 19/II (1896).
17. Paetzold, H. K.: Z. Aph. **32**, 303 (1953).
18. Bouška, J., u. Z. Švestka: BAC **2**, 6 (1950).
19. Link, F., u. Z. Linkova: BAC **5**, 82 (1954).
20. Schilling, G. F.: J. Atm. Sci. **22**, 110 (1965).
21. Nigam, R. C. J.: Geoph. Res. **69**, 1361 (1964).
22. Hemenway, C. L., D. S. Halgreen, and R. E. Coon: Space Res. **7**, 1423 (1967).
23. Link, F.: Bull. astron. **8**, 77 (1933).

1.6. Thermal Phenomena during Lunar Eclipses

1.6.1. Brief Outline of Temperature Measurements

Lunar temperature was first measured by Rosse in 1869 using a 3-foot reflector and two thermopiles placed in its focus. After some measurements made by Langley and another series by Very at the end of the last century, came Nicholson and Pettit in 1930 with their measurements at Mt. Wilson in the focus of the 100-inch mirror. They measured systematically the temperature of different spots on the lunar surface throughout the greater part of the lunation. To quote their principal result we give here the temperature of subsolar point at the Full Moon of about 400° K and at the center of the dark hemisphere of 120° K [1].

The lunar surface receives the radiation from the Sun and emits it in another domain of frequencies into space. The course of the temperature-phase curve depends on the thermal properties of the lunar surface i.e. thermal conductivity, specific heat and density. As during the eclipse the variation of incident solar radiation goes more than 200 times faster than during the lunation the temperature curve will be quite different. Moreover, detailed temperature measurements can disclose the thermal differences of several features on the lunar surface.

The infrared measurements of lunar radiation can be supplemented by microwave measurements. However, the execution of these determinations is more difficult.

It may be remembered that before the landing of the first space craft on the Moon the studies of thermal properties of the lunar surface played an important role in discussions us to whether or not there is dust on the Moon.

1.6.2. Theoretical Aspects of Surface Temperature Variations

In order to show the part played by lunar eclipses in the exploration of the thermal properties of the lunar surface, we shall give an outline of the theory according to the works of EPSTEIN [2], JAEGER [3] and WESSELINK [4].

The problem of thermal variations is given by the following equations where T is the temperature at any depth x below the surface of a homogeneous semi-infinite material having thermal conductivity k, density ρ and specific heat c [14].

The first is the classical heat conduction equation

$$\frac{\partial T}{\partial t} = \frac{k}{\rho c} \frac{\partial^2 T}{\partial x^2}. \tag{1.155}$$

The second for the heat flux directed outwards through the unity of surface is the definition of thermal conduction based on the analogy with the flow of liquids

$$F = k \frac{\partial T}{\partial x}. \tag{1.156}$$

At the surface this flux will be F_0, the temperature T_0 and hence, according to the Stephan-Boltzmann law, the emitted radiation is σT_0^4 and the radiation received from the Sun I. We have, therefore

$$\sigma T_0^4 = I + F_0. \tag{1.157}$$

For the received radiation I we must adopt the formula giving the flux, at any moment of the eclipse, calculated from the solar constant and the non-occulted portion of the solar disk taking into account the limb darkening in global radiation. This formula constitutes the fourth equation of our theory.

As the equation [Eq. (1.157)] for the boundary conditions is not linear, the solution cannot be found by making use of the Fourier series and some indirect method should be applied, such as numerical integration, used by WESSELINK, or the iterative process introduced by JAEGER which starts from the assumed solution and modifies it until it fits the observations.

In all these problems appearing as the parameter is the expression $\sqrt{k \rho c}$, called the *thermal inertia*, characterising the resistance of lunar material to temperature change: the greater its value, the smaller the temperature variation during an eclipse or other perturbations of thermal conditions. Ordinary rocks have inertia near 0,05, dry soil or sand 0.01 – 0.02 and pumice has the lowest known value 0.004, on the Earth.

1.6.3. Methods of Observing the Thermal Radiation of the Moon

The Moon's radiation reaching the Earth's surface has two distinct components:

a) Reflected sunlight between $0,3 - 5\,\mu$.

b) Infrared radiation emitted by the warmed lunar surface confined mainly between $8 - 14\,\mu$, i.e. in the transmission band of water vapour.

The radio waves observed by entirely different methods will be dealt with later. The small portion of the thermal radiation remaining between $0.3 - 8\,\mu$ is practically negligible in the lunar radiation flux.

The observation technique aims to separate both components and to measure their respective values. PETTIT and NICHOLSON made use of microscope cover-glass 0.165 mm thick which is superior to water-cell for the reasons of better transmission. The vacuum thermocouple was employed in the direct focus of the 100-inch telescope at Mt. Wilson, where the lunar image has the mean diameter of 116 mm. The width of the receiver on the thermocouple was about $0.62\,\text{cm}^2$, i.e. 0.55% of the lunar image. The cell of the vacuum thermocouple was closed by a rock-salt window, 2 mm thick, whose transmission is very high and constant over the whole measured spectral region.

It is beyond the scope of this book to give in detail the manner in which the measurements were made and reduced in order to give finally the temperature of the lunar surface. Ordinarily, besides the determination of drift curves before and after eclipse, the deflection on and off the lunar spot without filter and with the above filter were taken and, in addition, Arcturus and Vega were used as comparison stars.

The method of measuring one spot on the Moon was replaced by the scanning of the lunar surface developed by SINTON [5] and brought to high degree of perfection by SHORTHILL and SAARI. The last named authors have developed and elaborated a rapid scan system [6] for mapping the Moon in the visible ($0.45\,\mu$) and far infrared ($10 - 12\,\mu$) regions. Using a mercury-doped germanium detector cooled to liquid neon temperature they succeeded in scanning the lunar disk in 17 min at $10''$ resolution in about 200 traverses, the duration of each being about 5 sec. Both visible and infrared signals were recorded on magnetic tape and images were made from the scans by intensity modulating an osciloscope. On its screen the image was photographed line by line. For more details we refer to the above quoted paper.

1.6.4. Experimental Results

All results of eclipse measurements examples of which are given in Fig. 1.6.1 show without doubt two characteristic features:

i) Rapid progress of the cooling of lunar surface in the penumbra.

ii) Much slower decrease of the temperature in the umbra.

The penumbral behaviour is not compatible with the thermal inertia of solid rock and not even with any terrestrial material. The eclipse curve on October 27, 1939, observed by PETTIT, fits in the penumbral part according to the computations by WESSELINK with a thermal of inertia 0.001, while the lowest known terrestrial value is 0.004 for pumice. On the other hand, SMOLUCHOWSKI's [7] investigations have shown the decrease of thermal conductivity if powdered material is in the vacuum. Evidently the air in the interstices raises its value under normal pressure conditions. Adopting these results we obtain in high vacuum $\sqrt{k\rho c} = 0.001$, in good agreement with observations. This fact was the origin of the dust cover hypothesis generally accepted until not long ago.

In addition, JAEGER and HARPER [8] have proposed a variant to explain the slower cooling observed in the umbra as given by the simple layer theory (Fig. 1.6.1). This variant consists of a two-component model

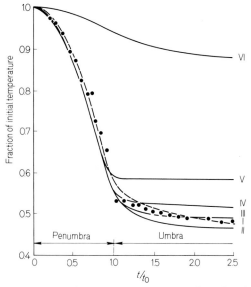

Fig. 1.6.1. Cooling of lunar surface during the lunar eclipse October 27, 1939 according to PETTIT (dots) and different models computed by JAEGER and HARPER (II – V) of decreasing thickness of the dust layer from 0.24 to 0.05 cm. The curve I computed with thermal inertia 0.00097 (only dust) and the curve VI with 0.05 (bare rock)

of the lunar surface with a layer of dust 1.7 mm thick having a thermal inertia of 0.00097 and overlying the solid base with inertia 0.01 (curve III, Fig. 1.6.1). In other words, during the totality when the solar radiation is absent the very low thermal inertia of dust could not explain the slow observed temperature decrease.

1.6.5. Hot Spots on the Moon

Let us specify that what is meant by hotness is the temperature difference, e.g. 216° K on the floor of the crater Tycho and 160° K in its vicinity. This phenomenon was discovered by SHORTHILL et al. [9] during the total eclipse of March 13, 1960 and confirmed by SINTON [5] during the eclipse of September 5, 1960.

Fig. 1.6.2 shows two pyrometric scans across Tycho illustrating the rapid temperature fall of the lunar surface, except for Tycho which happens in shadow to be warmer than its environs.

The method of scanning was considerably improved, as stated above, by SHORTHILL and SAARI [6] and has been applied recently to the eclipse of December 19, 1964 observed at Helwan Observatory in Egypt by

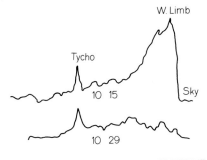

Fig. 1.6.2. Pyrometric scans across TYCHO during the eclipse on March 13, 1960 according to SINTON

SAARI et al. [6]. The results are presented in the form of blue or infrared images and for the Tycho region in the form of an isotherm map. Even when the principal interest lies in the selenological domain, we cannot resist the temptation to give here the main conclusions accompanied by a specimen image. On the image (Fig. 1.6.3) approximately 500 hot spots have been counted and about half of these have been identified or located on the lunar disk. Approximately 85% of then are the craters, the small ones showing greater temperature differences than the larger ones.

SINTON [5] explains this behaviour by a different thickness of dust covering the lunar surface. The vicinity of Tycho covered with a thicker layer of dust cools more rapidly than the floor of the crater where the dust is thinner. If it were really so, this might be an indication of the

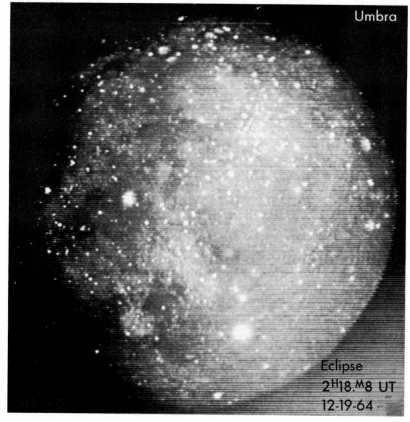

Fig. 1.6.3. Infrared scan of eclipsed Moon on December 19, 1964 showing the profusion of hot spots [6]

relative age, and the crater floor with a smaller dust deposit would be younger that the environs.

However Sinton's interpretation has not been universally accepted and some other hypotheses have been proposed. We have the choice of structural variations of lunar soil, of physical differences such as the infrared emissivity, thermal conduction or density and possibly also the source of internal heat. Future lunar explorers may, after all, provide us with the definitive answer.

1.6.6. Lunar Eclipse at Microwaves

As we go from the infrared to the mm-wave-length region, the penetration of rays below the surface increase. The surface layers of low thermal and electrical conductivity are partly transparent to microwaves, so that the radiation originated beneath the surface layers may be observed. In fact, this is what happens during the lunar cycle and naturally during the eclipses too, though to a different extent.

During the lunation the infrared temperature varies from about $100-400°$ K. That is also the variation of the thin surface layer. The same kind of variation but at 12.5 mm shows an amplitude of only $100°$ K with a lag of 1/8 lunation [10].

During eclipses, where the solar radiation varies even more rapidly, we may expect a small temperature amplitude accompanied by a considerable lag behind the optical phase. Sinton's measurements [11] at 1.5 mm of the eclipse of January 18, 1954 fulfilled this expectation. Only a $100°$ decrease was found with a lag of about 1 hour. GIBSON [12] on the other hand observing at 8.6 mm wave found only a decrease of $1°$ during the eclipse of March 13, 1960. At the same eclipse CASTELLI et al. observing at 10 and 23 cm wave length found no variation greater than 2.5% [13].

These and other similar findings lead not only to several models of the lunar surface [14] but give us the possibility of estimating the numerical value of electrical conductivity approaching that of dry sandy soil. Our feeling is, that in the near future we shall have the opportunity to verify or reject the most ingenious speculations elaborated so far on the lunar surface not only by walking on it, but to examine finally so-called "lunite" by our terrestrial methods.

Bibliography

1. PETTIT, E., and S. B. NICHOLSON: Aph. J. **71**, 102 (1930).
2. EPSTEIN, P. S.: Phys. Rev. **33**, 269 (1929).
3. JAEGER, J. C.: Australian J. Phys. **6**, 10 (1953).
4. WESSELINK, A. J.: Bull. Astron. Inst. Neth. **10**, 356 (1948).
5. SINTON, W. M.: Lowell Obs. Bull. 108, 25 (1960).
6. SAARI, J. M., R. W. SHORTHILL, and T. K. DEATON: Boeing Docum. D1-82-0533 (1966).
7. SMOLUCHOWSKI, M.: Bull. Acad. Sci. Cracovie A, 129 (1910); A, 548 (1911).
8. JAEGER, J. C., and A. F. HARPER: Nature **166**, 1026 (1950).
9. SHORTHILL, R. W., H. C. BOROUGH, and J. M. CONLEY: Publs. Astron. Soc. Pacific **72**, 481 (1960).
10. PIDDINGTON, J. H., and H. C. MINNET: Australian J. Sci. Research, Ser. A**2**, 63 (1949).
11. SINTON, W. M.: Aph. J. **123**, 325 (1965).
12. GIBSON, J. E.: Aph. J. **133**, 1072 (1961).
13. CASTELLI, J. P., C. P. FERIOLI, and J. AARONS: Astr. J. **65**, 485 (1960).
14. SINTON, W. M.: Temperature on the lunar surface, Physics and Astronomy of the Moon (ed. KOPAL), p. 407—427. New York-London 1962.

2. Eclipses of Artificial Earth Satellites

2.1. Preliminary Remarks

New possibilities have been opened up by recent developments in the field of astronautics, enabling us to inject in different orbits artificial satellites of the Earth. In principle there are two distinct categories of satellites operating in the eclipse domain:

A. *Passive satellites* in the form of huge balloons reflecting the solar radiation towards the Earth's surface where photometrical measurements can be performed if the geometrical and meteorological conditions are favourable.

B. *Active satellites* furnished with photometrical equipment and telemetering their indications to the terrestrial tracking stations.

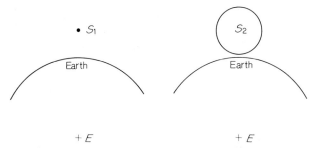

Fig. 2.1. Aspects of the Sun from the satellite (left) and the Moon (right) in respect of the Earth

Over the lunar eclipses we have many advantages accompanied, however, by some minor disadvantages. Among the former we can count:

a) Very great frequency of satellite eclipses — several tens per year — in comparison with 2–3 possible eclipses in the same time interval.

b) Great selectivity in the altitude, the light beam having in the upper atmosphere a vertical extension smaller than for lunar eclipses (Fig. 2.1).

c) Great selectivity in latitude, as the effective part of the shadow terminator is very small (<100 km) (Fig. 2.1).

d) Possibility to explore the short wave radiations with active satellites.

Nevertheless, some disadvantages or difficulties of satellite eclipses should be mentioned in all objectivity:

e) Necessity of an exact ephemeris. This problem was solved a long time ago for the Moon but remains outstanding especially for balloon satellites whose motion is capricious.

f) Irregular shape of the balloon satellites in combination with their rotation sometimes deforms the light curve during the eclipse.

g) Necessity of tracking the satellite. This operation is more complicated than for the slow-moving Moon.

h) Irregularities or perturbations affecting the complicated apparatus of active satellites.

i) High costs of constructing an artificial satellite. However, the satellite once injected with success in the orbit may serve, especially in passive category, over a period of several years.

2.2. General Conditions of Visibility

Instead of geometrical conditions, we consider here at once actual conditions given by the formulae for the radius of the umbra (Fig. 2.2)

$$\sigma = \pi_\odot + \pi_S - R_\odot - \omega_0 \tag{2.1}$$

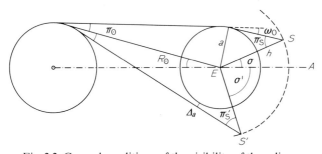

Fig. 2.2. General conditions of the visibility of the eclipse

and of the penumbra

$$\sigma' = \pi_\odot + \pi_S + R_\odot$$

where the parallax π_S of the satellite at the height h is

$$\sin \pi_S = \frac{a}{a+h}; \quad \sin \pi'_S = \frac{a+\Delta a}{a+h}. \tag{2.2}$$

We include in these formulae the increase of the Earth's radius $\Delta a = 2-3\%$ and the refraction, on the average $\omega_0 = 1{,}2°$. At the moment when the angular distance of the satellite from the antisun reaches σ_1, its brightness begins to decline and at the distance σ_0 the satellite disappears.

2.3. Ephemeris of the Eclipse

Large variations of the altitude h make the computation of the eclipse more difficult, while the parallaxes and hence both radii σ and σ' are variable during the eclipse. Nevertheless we may assume the linearity of these variations during the phenomenon.

We shall give here a method for the prediction of eclipses [1] based on the circulars issued bimonthly by the Astrophysical Observatory of the Smithsonian Institution, Cambridge, Mass. We find there the positions of the subsatellite points at different parallels δ with cor-

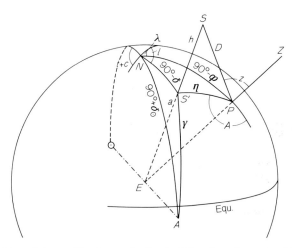

Fig. 2.3. Position of the satellite S near the Earth, P the observing point, N the North pole, $\odot A$ the direction from the Sun to the antisun

responding altitudes h of the satellite. For the computation we take three subsatellite points lying in the vicinity of our observing station P (Fig. 2.3) and we compute for every point the angular distance of the satellite from the shadow axis or from the antisolar point A.

In the spherical triangle: north pole N, subsatellite point S' and the antisun A (Fig. 2.3) the angular distance γ is

$$\cos \gamma = -\sin \delta_\odot \sin \delta + \cos \delta_\odot \cos \delta \cos(t+C-l) \qquad (2.3)$$

where $\delta_\odot, \delta$ are the declinations of the Sun and the point S',
   t the moment of the passage through δ parallel expressed in UT,
   C the time-equation, i.e. true-apparent time,
   l the longitude of S'-positive from Greenwich towards E.

For the same moment t we compute also the radii σ and σ'. We represent further γ, σ, σ' as the function of the time t (Fig. 2.4), and we

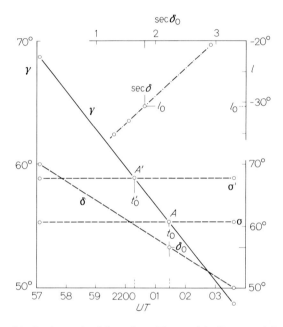

Fig. 2.4. Graphical ephemeris of the eclipse. The straight lines γ and δ give the variation as a function of the time (UT). Above the longitude l_0 as a function of sec δ

generally obtain three straight lines (or nearly so) whose intersections A, A' give the moments t_0 and t'_0 for the contacts with the umbra and the penumbra.

For the topocentrical positions of the satellite we use the spherical triangle N, S', P. We calculate first the geocentrical angle

$$\cos \eta = \sin \delta \sin \varphi + \cos \delta \cos \varphi \cos(l - \lambda) \qquad (2.4)$$

from which we get the azimuth

$$\sin A = \frac{\sin(l - \lambda)}{\sin \eta} \cos \delta \qquad (2.5)$$

and finally the zenithal distance

$$\cos \beta = \frac{h}{2a+h} \operatorname{ctg} \frac{\eta}{2}$$
$$z = \beta + \frac{\eta}{2}. \tag{2.6}$$

The linear distance $PS = r$ is then

$$r = \frac{a+h}{\sin z} \sin \eta. \tag{2.7}$$

The solution of Eqs. (2.5)–(2.7) may be performed with aid of tables by ŽONGOLOVIČ and AMELIN [2].

Using the same graphical method (Fig. 2.4) we determine the declination δ_0 of the subsatellite point or its latitude at the moment of the eclipse from the graph of δ as the function of t. For the longitude l_0 of this point it is preferable to trace l as the function of sec δ (Fig. 2.4).

During the year the conditions of visibility change with the changing position of the satellite orbit with regard to the antisun. Let us consider the celestical sphere (Fig. 2.5) with the satellite orbit projected on it. We shall compute the orbital coordinates of the antisun related to the orbital plane.

$$\sin B = -\sin \delta_\odot \cos i + \cos \delta_\odot \sin i \sin(\alpha_\odot - \Omega)$$
$$\sin L = -\cos(\alpha_\odot - N) \frac{\cos \delta_\odot}{\cos B} \tag{2.8}$$

where $\alpha_\odot, \delta_\odot$ the equatorial coordinates of the Sun, i the inclination of the orbit towards the equator and N the right ascension of the node.

The series of eclipses begins at the moment when the orbital latitude B of the antisun attains the value of $\pm \sigma_0$ and ends at the value $\mp \sigma_0$ (Fig. 2.5). In the intermediary position of the antisun A there are two limiting latitudes of the subsatellite points, one for the entry and the other for the exit of the shadow

$$\sin \varphi_{N,S} = \sin i \cos(L \mp m)$$
$$\cos m = \frac{\cos \sigma_0}{\cos B}. \tag{2.9}$$

These limits are useful for the preliminary determination of observing conditions. We compare our latitude with the above values of $\varphi_{N,S}$. If they are not too different, we may proceed to the exact computation of the eclipse ephemeris.

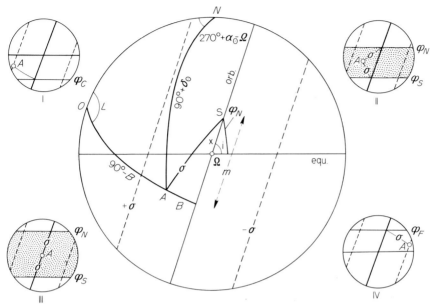

Fig. 2.5. General conditions of the visibility of the eclipse represented on the sphere. S the satellite, A the antisun, N the North pole, σ the radius of the umbra. Successive phases of the visibility: I the antisun reaches the parallel $+\sigma$, i.e. the beginning of the eclipse series, II the antisun approaches to the orbital plane, III the maximum latitude extension of the eclipse, IV the antisun leaves the parallel $-\sigma$ i.e. the end of the eclipse series

The exact ephemeris necessary for the reduction of the observations, i.e. the angular distance γ as a function of the time t, may be computed from the exact orbital elements if they are available. The classical formulae of celestial mechanics give with the usual notation:

$$M = E - e \sin E \qquad \mathrm{tg}(\alpha - \Omega) = \mathrm{tg}\, x \cos i$$

$$\mathrm{tg}\frac{v}{2} = \mathrm{tg}\frac{E}{2}\sqrt{\frac{1+e}{1-e}} \qquad a+h = \frac{\mathrm{perig.}}{1-e}(1 - e \cos E) \qquad (2.10)$$

$$x = \omega + v$$

$$\sin \delta = \sin x \sin i$$

and with Eq. (2.3) we obtain γ.

Recently, the above mentioned Circulars SAO began to give for every day the average value of x_0 for entry and exit of the umbra. It is, therefore, possible to determine the corresponding time t_0 using the graph of x as the function of t.

2.4. Photometrical Theory of the Shadow [3]

The scene of satellite eclipses is analogous to that of lunar eclipses (Fig. 2.6) with some modifications due to the different scales of the two phenomena. The light element in the solar plane illuminating the point N

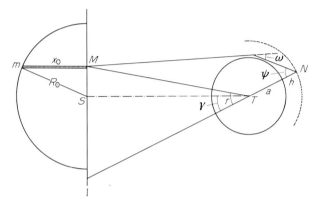

Fig. 2.6. Scene of the satellite eclipse

on the satellite orbit is now the strip s, while the point E is in reality very far from S. The light intensity of the strip may be obtained as a limiting case of Eq. (1.17) for $r \to \gamma \to \infty$ and $m \to \varepsilon_0$

$$di = 2\left[(1-\kappa)x_0 + \frac{\pi\kappa}{4R_\odot}x_0^2\right]dr \qquad (2.11)$$

$$r\varepsilon_0 = x_0 = \sqrt{R_\odot^2 - (r-\gamma)^2}.$$

If we now write

$$P = 2x_0, \quad Q = P - \frac{\pi x_0^2}{2R_\odot}$$

we get a similar expressions as for lunar eclipses namely

$$di = (P - \kappa Q)\,dr \qquad (2.12)$$

with the numerical values of P and Q contained in the Table 2.1.
The illumination at N is then

$$e = k \int_{\gamma - R_\odot}^{\gamma + R_\odot} T(h_0)\,di \qquad (2.13)$$

where the general transmission coefficient $T(h_0)$ has its usual components

$$T(h_0) = 10^{-AM}\,\Phi^{-1}$$

namely the extinction 10^{-AM} and the attenuation by refraction Φ.

Table 2.1

No	r-γ (°)	P	Q	No.	r-γ (°)	P	Q
1	0.000	32.00	6.87	9	0.133	27.72	8.87
2	0.017	31.94	6.91	10	0.150	26.46	9.28
3	0.033	31.74	7.00	11	0.167	24.98	9.66
4	0.050	31.44	7.19	12	0.183	23.24	9.99
5	0.067	30.98	7.42	13	0.200	21.16	10.16
6	0.083	30.40	7.72	14	0.217	18.66	10.12
7	0.100	29.66	8.06	15	0.233	15.48	9.59
8	0.117	28.78	8.46	16	0.250	11.14	8.10

2.5. General Transmission Coefficient

Because of a different scale we must derive a new expression for the attenuation by refraction. Let us consider (Fig. 2.7) the radiation flux emitted from M and contained between two cones c and c' lying close to each other. In the absence of refraction the flux would reach the

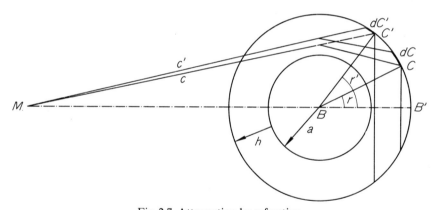

Fig. 2.7. Attenuation by refraction

sphere of the satellite in the annular zone dC'. Because of the refraction the illuminated annular zone will be dC. The attenuation will be as usual the ratio

$$\Phi = \frac{dC}{dC'}. \tag{2.14}$$

In order to determine it we compute first the spherical caps C and C'

$$C = 2\pi(a+h)^2 (1-\cos r)$$
$$C' = 2\pi(a+h)^2 (1-\cos r') \tag{2.15}$$

---

9 Link, Eclipse Phenomena

where

$$r = \pi_\odot \left(1 + \frac{h'_0}{a}\right) + \psi - \omega$$

$$r' = \pi_\odot \left(1 + \frac{h'_0}{a}\right) + \psi \qquad (2.16)$$

$$\psi = \arcsin \frac{a + h'_0}{a + h}$$

and we obtain by differentiation

$$dC = 2\pi (a+h)^2 \sin r \left[\frac{\pi_\odot}{a} + \frac{1}{(a+h)\cos\psi} - \frac{d\omega}{dh'_0}\right] dh'_0$$

$$dC' = 2\pi (a+h)^2 \sin r' \left[\frac{\pi_\odot}{a} + \frac{1}{(a+h)\cos\psi}\right] dh'_0 \qquad (2.17)$$

which finally gives the wanted ratio

$$\Phi = \frac{\sin\left[\pi_\odot \left(1 + \frac{h'_0}{a}\right) + \psi - \omega\right]}{\sin\left[\pi_\odot \left(1 + \frac{h'_0}{a}\right) + \psi\right]} \left[1 - \frac{a(a+h)\cos\psi \frac{d\omega}{dh'_0}}{\pi_\odot (a+h)\cos\psi + a}\right] \qquad (2.18)$$

This very general expression may be simplified by putting $\pi_\odot = 0$ and we get

$$\Phi = \frac{\sin(\psi - \omega)}{\sin \psi} \left[1 - (a+h)\cos\psi \frac{d\omega}{dh'_0}\right]. \qquad (2.19)$$

The general transmission coefficient is then [Eq. (1.36 b)]

$$-\log T = D' = AM + \log \Phi \qquad (1.36\,\mathrm{b})$$

and is nothing but the density of the auxiliary shadow (1.2.11).

Outside the shadow we have as usual

$$E = k \int_{\gamma - R_\odot}^{\gamma + R_\odot} di = \pi R_\odot^2 \left(1 - \frac{\kappa}{3}\right) \qquad (1.13)$$

and the density of the shadow is given by

$$D = \log_{10} E - \log_{10} e. \qquad (1.14)$$

The general form of some illumination integrals is shown in Fig. 2.8. The selectivity in the height increases with the decreasing altitude of the satellite. In comparison with lunar eclipses, there is more selectivity above 20 km which will be useful in the ozone investigations.

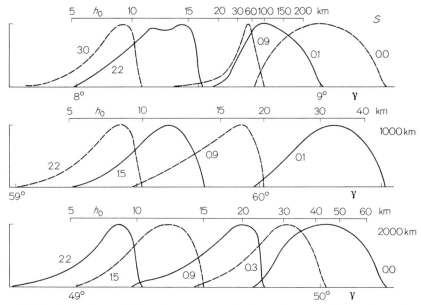

Fig. 2.8. Some forms of illumination integrals: S stationary satellite, and at the heights of 2000 and 1000 km

2.6. Simplified Presentation of the Eclipse Theory [4]

The density of the shadow is not only the function of the angular distance γ from the shadow axis but depends at least on three parameters, too:

a) The aerological situation on the shadow terminator i.e. the latitude and the season.

b) The height of the satellite.

c) The spectral band used.

These parameters are variable within wide limits and the tables taking these into account would be necessarily very extensive and owing to the interpolations not easy for practical use.

We looked therefore for an economical presentation of tables and the following method has been found the most convenient for this purpose. With the aid of the electronic computer, we obtain a complete and very extensive collection of tables and we make from it a summary satisfying simultaneously the commodity of use and the economy of presentation. To arrive at this goal we must restrict the number of parameters and the simplified theory of eclipses might guide us in this task.

Table 2.2

D	Situation									
	1	16	2	17	3	18	4	19	5	20

Red (6200 Å) – 900 km

D	1	16	2	17	3	18	4	19	5	20
0.05	+57 / +59	+58 / +60	+57 / +59	+58 / +58	+59 / +59	+59 / +60	+60 / +60	+61 / +61	+59 / +58	+60 / +60
0.1	+48 / +49	+49 / +50	+48 / +49	+48 / +49	+49 / +50	+50 / +50	+50 / +50	+51 / +51	+49 / +49	+50 / +50
0.15	+41 / +41	+42 / +42	+41 / +43	+42 / +42	+42 / +43	+43 / +44	+43 / +43	+44 / +44	+43 / +43	+44 / +44
0.2	+35 / +36	+36 / +37	+35 / +37	+36 / +36	+36 / +37	+37 / +38	+37 / +38	+38 / +39	+37 / +37	+38 / +39
0.4	+19 / +19	+20 / +21	+19 / +21	+20 / +19	+19 / +21	+21 / +22	+20 / +22	+22 / +23	+20 / +21	+22 / +23
0.6	+05 / +06	+08 / +08	+05 / +08	+07 / +06	+06 / +07	+08 / +10	+07 / +08	+09 / +11	+05 / +07	+08 / +10
0.8	−07 / −05	−04 / −03	−07 / −03	−04 / −06	−06 / −06	−03 / −00	−05 / −09	−02 / −03	−10 / −10	−06 / −05
1.0	−19 / −17	−15 / −13	−20 / −14	−15 / −19	−20 / −21	−14 / −10	−19 / −25	−14 / −18	−24 / −25	−18 / −19
1.2	−34 / −34	−27 / −25	−35 / −27	−28 / −36	−36 / −37	−28 / −22	−35 / −36	−27 / −30	−36 / −37	−30 / −30
1.4	−50 / −51	−40 / −41	−50 / −42	−41 / −51	−50 / −49	−41 / −34	−49 / −46	−40 / −39	−48 / −47	−40 / −40
1.6	−63 / −61	−53 / −53	−62 / −54	−53 / −62	−61 / −59	−52 / −45	−60 / −56	−51 / −48	−59 / −57	−50 / −49
1.8	−74 / −72	−64 / −61	−72 / −62	−63 / −71	−71 / −70	−62 / −54	−70 / —	−60 / −57	−70 / —	−59 / −57
2.0	−83 / —	−74 / −71	−83 / −70	−72 / −82	−82 / —	−70 / −62	−81 / —	−71 / −66	−81 / —	−68 / —

Simplified Presentation of the Eclipse Theory

Red (6200 Å) - 1000 km															
0.05	+55/+56	+55/+57	+55/+56	+54/+55	+55/+56	+55/+57	+57/+57	+57/+57	+57/+58	+58/+58	+59/+59	+59/+59	+57/+56	+57/+57	+58/+57
0.1	+46/+46	+46/+47	+47/+48	+45/+46	+46/+47	+46/+47	+47/+47	+48/+48	+48/+49	+49/+49	+49/+49	+50/+50	+48/+47	+48/+48	+49/+48
0.15	+39/+39	+39/+40	+39/+40	+39/+39	+39/+40	+40/+41	+40/+41	+41/+42	+42/+42	+42/+42	+43/+43	+43/+44	+41/+41	+42/+42	+43/+43
0.2	+33/+34	+34/+35	+35/+36	+33/+34	+34/+35	+35/+36	+34/+36	+35/+36	+36/+37	+36/+37	+37/+38	+37/+38	+36/+36	+37/+37	+37/+38
0.4	+17/+18	+19/+19	+20/+20	+17/+18	+18/+19	+20/+20	+18/+20	+20/+21	+21/+22	+20/+22	+21/+23	+23/+24	+19/+20	+21/+22	+22/+23
0.6	+05/+05	+07/+07	+08/+09	+04/+05	+06/+07	+08/+09	+06/+07	+08/+09	+09/+11	+07/+08	+09/+10	+11/+12	+05/+07	+08/+09	+10/+11
0.8	−07/−06	−04/−03	−02/−01	−08/−07	−05/−04	−03/−01	−06/−06	−03/−02	−01/+00	−05/−08	−02/−03	+01/+01	−10/−09	−05/−05	−02/−01
1.0	−19/−18	−15/−14	−15/−10	−21/−20	−16/−15	−13/−11	−20/−21	−14/−15	−11/−11	−18/−24	−13/−17	−09/−12	−24/−24	−18/−19	−14/−14
1.2	−34/−33	−27/−25	−22/−20	−35/−36	−28/−27	−23/−22	−35/−36	−28/−28	−22/−22	−34/−35	−26/−29	−20/−24	−35/−36	−29/−30	−24/−25
1.4	−49/−51	−40/−40	−33/−31	−50/−52	−41/−42	−34/−34	−49/−49	−41/−41	−34/−34	−48/−45	−39/−39	−32/−33	−47/−46	−39/−40	−33/−34
1.6	−62/−61	−53/−53	−44/−45	−62/−62	−53/−54	−45/−46	−61/−58	−52/−51	−45/−44	−59/−55	−50/−47	−43/−41	−58/−56	−49/−48	−42/−42
1.8	−74/−72	−64/−62	−55/−55	−72/−71	−63/−63	−56/−56	−70/−69	−61/−59	−54/−52	−69/−67	−59/−56	−52/−49	−68/—	−59/−57	−51/−49
2.0	−83/−83	−74/−71	−65/−62	−82/−81	−72/−71	−64/−63	−81/—	−70/−67	−62/−59	−79/—	−68/−65	−60/−56	−80/—	−67/—	−59/−57

Table 2.2 (Continued)

D	Situation									
	1	16	2	17	3	18	4	19	5	20

Red (6200 Å) - 1100 km

D	1	16	2	17	3	18	4	19	5	20
0.05	+53/+55	+54/+55	+54/+54	+55/+55	+55/+55	+55/+55	+54/+54	+55/+55	+55/+54	+55/+55
0.1	+44/+45	+45/+46	+45/+45	+46/+47	+45/+46	+46/+47	+45/+46	+46/+46	+45/+45	+46/+46
0.15	+38/+38	+38/+39	+38/+39	+39/+40	+38/+39	+39/+40	+39/+39	+40/+40	+39/+39	+40/+40
0.2	+33/+33	+33/+34	+33/+34	+34/+35	+33/+34	+34/+35	+34/+34	+34/+35	+34/+34	+35/+35
0.4	+17/+17	+18/+19	+17/+18	+19/+19	+17/+19	+19/+20	+18/+19	+19/+20	+18/+19	+20/+21
0.6	+05/+05	+06/+07	+05/+06	+07/+08	+05/+06	+07/+08	+05/+06	+07/+08	+04/+06	+08/+10
0.8	−07/−05	−04/−03	−07/−06	−04/−03	−06/−06	−04/−03	−06/−09	−03/−04	−10/−10	−06/−05	−02/−02
1.0	−19/−17	−15/−13	−19/−18	−15/−14	−19/−20	−14/−15	−19/−25	−14/−18	−24/−25	−19/−19	−14/−14
1.2	−32/−32	−26/−24	−33/−34	−26/−26	−35/−36	−27/−28	−34/−36	−27/−30	−35/−37	−30/−30	−25/−25
1.4	−48/−50	−39/−39	−48/−50	−39/−40	−40/−49	−40/−40	−49/−46	−40/−40	−47/−47	−40/−40	−34/−35
1.6	−61/−60	−51/−52	−61/−60	−52/−52	−60/−58	−52/−51	−59/−56	−51/−48	−58/−56	−50/−48	−43/−43
1.8	−73/−71	−63/−61	−70/−70	−62/−61	−70/−68	−61/−59	−70/−67	−60/−57	−68/—	−59/−57	−52/−50
2.0	−82/−81	−73/−70	−80/—	−70/−69	−80/—	−70/−67	−80/—	−69/−66	−80/—	−67/—	−59/−57

Simplified Presentation of the Eclipse Theory

Red (6200 Å) - 1200 km

0.05	+52 / 53	+52 / 53	+52 / 54	+51 / 52	+52 / 52	+52 / 53	+52 / 53	+53 / 53	+53 / 53	+53 / 53	+54 / 54	+53 / 52	+53 / 53	+54 / 53
0.1	+43 / 44	+44 / 44	+44 / 45	+42 / 43	+43 / 44	+43 / 44	+44 / 45	+44 / 45	+44 / 44	+44 / 45	+45 / 45	+44 / 44	+45 / 44	+45 / 45
0.15	+37 / 37	+37 / 38	+38 / 39	+36 / 37	+37 / 37	+37 / 38	+38 / 38	+38 / 39	+38 / 39	+38 / 39	+39 / 40	+38 / 38	+39 / 38	+39 / 39
0.2	+32 / 32	+32 / 33	+33 / 34	+31 / 32	+32 / 32	+32 / 33	+32 / 33	+33 / 34	+33 / 34	+33 / 34	+34 / 35	+33 / 33	+34 / 34	+34 / 34
0.4	+16 / 17	+18 / 18	+19 / 19	+16 / 16	+17 / 17	+17 / 19	+18 / 17	+19 / 20	+17 / 19	+17 / 19	+20 / 21	+17 / 18	+19 / 19	+20 / 20
0.6	+04 / 05	+06 / 07	+08 / 08	+03 / 04	+05 / 06	+07 / 08	+06 / 07	+08 / 09	+04 / 06	+05 / 06	+09 / 10	+04 / 06	+06 / 08	+08 / 09
0.8	−07 / 06	−04 / 03	−02 / 01	−08 / 07	−05 / 04	−03 / 02	−04 / 03	−02 / 01	−07 / 06	−06 / 08	−01 / 01	−03 / 04	−06 / 05	−03 / 02
1.0	−18 / 17	−15 / 13	−12 / 10	−20 / 19	−16 / 15	−13 / 12	−15 / 15	−11 / 11	−19 / 20	−18 / 24	−11 / 13	−14 / 18	−18 / 19	−14 / 14
1.2	−32 / 30	−25 / 24	−21 / 20	−34 / 34	−27 / 26	−22 / 21	−27 / 28	−22 / 22	−34 / 35	−33 / 36	−21 / 25	−26 / 30	−29 / 30	−25 / 25
1.4	−47 / 49	−38 / 38	−32 / 30	−49 / 50	−40 / 40	−33 / 33	−40 / 40	−33 / 33	−48 / 48	−48 / 46	−32 / 34	−39 / 39	−39 / 40	−34 / 34
1.6	−60 / 60	−51 / 52	−43 / 43	−61 / 61	−52 / 53	−44 / 45	−51 / 51	−44 / 44	−60 / 58	−59 / 55	−43 / 42	−50 / 48	−49 / 48	−42 / 42
1.8	−72 / 70	−62 / 61	−54 / 54	−71 / 70	−62 / 61	−55 / 55	−61 / 59	−54 / 52	−70 / 67	−68 / 66	−53 / 49	−60 / 56	−58 / 56	−51 / 50
2.0	−81 / 80	−72 / 69	−63 / 61	−81 / 80	−71 / 70	−63 / 62	−69 / 67	−62 / 59	−80 / —	−79 / —	−60 / 56	−68 / 65	−67 / —	−59 / 57

136 Eclipses of Artificial Earth Satellites

Table 2.2 (Continued)

D	Situation									
	1	2	3	4	5					
	16	17	18	19	20					

Blue (4600 Å) - 900 km

D	1		2		3		4		5	
	16		17		18		19		20	
0.05	+61	+62	+62	+63	+63	+64	+64	+66	+64	+65
	63	65	64	65	64	65	64	66	63	65
0.1	+52	+54	+53	+55	+54	+56	+55	+57	+55	+56
	55	56	55	56	55	57	55	57	54	56
0.15	+47	+48	+47	+49	+48	+50	+49	+51	+48	+50
	49	50	48	50	49	51	49	51	46	50
0.2	+42	+44	+42	+44	+43	+45	+44	+46	+44	+46
	43	46	44	46	44	46	45	47	44	46
0.4	+28	+30	+28	+31	+29	+32	+30	+33	+30	+32
	29	32	29	32	30	33	31	33	30	32
0.6	+18	+21	+18	+21	+19	+22	+20	+23	+20	+23
	19	25	20	23	21	23	21	24	20	23
0.8	+10	+14	+10	+14	+11	+15	+12	+16	+12	+15
	11	15	11	15	12	16	13	17	12	16
1.0	+03	+07	+03	+07	+04	+08	+05	+09	+04	+08
	04	08	04	08	05	09	06	10	05	09
1.2	−04	+01	−04	+01	−03	+02	−02	+03	−03	+02
	02	06	03	02	02	03	01	04	03	03
1.4	−10	−05	−10	−05	−09	−03	−08	−02	−10	−04
	09	+01	09	03	08	02	09	02	10	03
1.6	−16	−10	−16	−10	−15	−08	−14	−08	−17	−10
	15	04	15	09	15	03	15	08	17	09
1.8	−22	−15	−22	−15	−22	−14	−20	−13	−23	−15
	20	09	21	14	21	13	22	14	23	15
2.0	−29	−20	−28	−20	−28	−19	−27	−18	−28	−20
	26	13	27	19	27	18	28	19	28	20

Simplified Presentation of the Eclipse Theory

Blue (4600 Å) - 1000 km													
0.05	+58/+60	+59/+62	+60/+63	+60/+62	+61/+62	+60/+61	+61/+62	+62/+62	+63/+63	+64/+64	+61/+61	+62/+62	+64/+63
0.1	+50/+52	+51/+53	+52/+55	+51/+53	+52/+54	+52/+53	+53/+54	+53/+54	+55/+55	+56/+56	+53/+52	+54/+54	+55/+55
0.15	+44/+46	+46/+48	+47/+49	+46/+47	+47/+49	+46/+47	+48/+48	+48/+48	+49/+50	+51/+51	+47/+46	+48/+48	+50/+49
0.2	+39/+41	+41/+43	+43/+45	+41/+43	+43/+44	+41/+42	+43/+44	+43/+43	+45/+45	+46/+47	+42/+42	+44/+44	+45/+45
0.4	+26/+27	+28/+30	+30/+32	+28/+30	+30/+32	+28/+29	+30/+31	+29/+30	+32/+32	+34/+34	+29/+29	+31/+31	+33/+33
0.6	+17/+18	+19/+21	+21/+23	+19/+20	+21/+23	+18/+19	+21/+22	+20/+21	+22/+23	+25/+25	+19/+19	+22/+22	+24/+24
0.8	+09/+10	+12/+13	+14/+16	+12/+13	+14/+16	+10/+11	+13/+15	+12/+13	+15/+16	+18/+18	+11/+11	+14/+15	+17/+17
1.0	+02/+03	+05/+07	+08/+10	+05/+07	+08/+10	+03/+04	+07/+08	+05/+05	+09/+09	+12/+12	+03/+04	+08/+08	+11/+11
1.2	−05/−04	00/−01	+03/+04	−01/+01	+03/+04	−03/−02	+01/+02	−02/−02	+03/+03	+06/+07	−04/−03	+01/+02	+05/+06
1.4	−11/−10	−06/−04	−02/−01	−06/−05	−02/−01	−09/−09	−04/−03	−08/−09	−03/−02	+01/+02	−11/−10	−05/−04	00/00
1.6	−17/−16	−11/−10	−07/−05	−11/−10	−07/−06	−16/−15	−09/−09	−14/−15	−08/−08	−04/−03	−17/−17	−10/−10	−05/−05
1.8	−23/−21	−16/−15	−11/−10	−16/−15	−11/−10	−22/−21	−14/−14	−20/−22	−13/−14	−08/−08	−23/−23	−15/−15	−10/−10
2.0	−29/−27	−21/−19	−15/−14	−21/−20	−16/−14	−28/−27	−20/−19	−26/−27	−18/−19	−12/−13	−28/−28	−20/−20	−14/−15

138   Eclipses of Artificial Earth Satellites

Table 2.2 (Continued)

D	Situation				
	1 16	2 17	3 18	4 19	5 20
Blue (4600 Å) — 1100 km					
0.05	+56 +57 +58 +58 +60 +61	+57 +59 +60 +59 +60 +61	+58 +59 +60 +58 +60 +61	+59 +60 +60 +59 +60 +61	+58 +59 +60 +57 +59 +60
0.1	+48 +49 +51 +50 +52 +53	+49 +51 +52 +51 +52 +53	+50 +51 +52 +50 +52 +53	+50 +51 +53 +50 +51 +53	+50 +51 +52 +49 +53 +52
0.15	+43 +44 +46 +44 +46 +47	+43 +45 +46 +45 +46 +48	+44 +46 +47 +45 +46 +48	+44 +46 +47 +44 +46 +48	+44 +46 +47 +44 +45 +47
0.2	+38 +40 +41 +40 +42 +43	+39 +41 +42 +40 +42 +43	+39 +41 +43 +40 +42 +44	+40 +41 +43 +40 +42 +43	+40 +41 +43 +39 +41 +43
0.4	+25 +27 +29 +26 +29 +31	+26 +28 +30 +27 +29 +31	+25 +27 +29 +27 +29 +31	+27 +29 +31 +27 +29 +31	+26 +29 +30 +27 +29 +30
0.6	+16 +18 +20 +17 +20 +22	+17 +19 +21 +18 +20 +23	+16 +18 +20 +18 +20 +23	+17 +20 +22 +18 +20 +22	+17 +20 +22 +17 +20 +22
0.8	+08 +11 +14 +09 +12 +15	+09 +12 +14 +10 +13 +16	+08 +11 +14 +10 +13 +16	+09 +13 +15 +10 +13 +16	+09 +12 +15 +10 +13 +15
1.0	+01 +05 +08 +02 +06 +09	+02 +06 +08 +03 +07 +10	+01 +05 +08 +03 +07 +10	+02 +06 +09 +03 +07 +10	+02 +06 +09 +03 +06 +09
1.2	−05 −01 +02 −04 +01 +04	−05 00 +03 −04 +01 +04	−05 −01 +02 −03 +01 +04	−04 00 +04 −04 +01 +04	−05 00 +03 −05 00 +04
1.4	−11 −06 −03 −10 −05 −01	−11 −06 −02 −10 −04 −01	−11 −06 −03 −10 −04 00	−10 −05 −01 −11 −05 −01	−12 −06 −02 −12 −05 −01
1.6	−17 −11 −07 −16 −10 −06	−17 −11 −06 −15 −10 −05	−17 −11 −07 −16 −09 −05	−16 −10 −06 −17 −10 −05	−18 −11 −07 −18 −11 −06
1.8	−23 −16 −11 −21 −15 −10	−23 −16 −11 −21 −14 −10	−23 −16 −11 −22 −15 −10	−22 −15 −10 −23 −16 −10	−24 −16 −11 −24 −17 −11
2.0	−29 −21 −16 −27 −20 −14	−28 −21 −15 −27 −19 −14	−29 −21 −16 −28 −20 −14	−28 −20 −15 −29 −21 −15	−29 −22 −16 −29 −22 −16

Blue (4600 Å) — 1200 km

0.05	+55 +57	+56 +58	+57 +58	+54 +55	+55 +57	+56 +58	+56 +56	+57 +57	+57 +58	+57 +57	+58 +58	+59 +59	+56 +56	+57 +57	+58 +58
0.1	+47 +48	+48 +50	+49 +51	+46 +48	+48 +49	+49 +51	+48 +48	+49 +50	+50 +51	+49 +49	+50 +50	+51 +51	+48 +48	+49 +49	+50 +50
0.15	+41 +43	+43 +45	+44 +46	+41 +42	+42 +44	+43 +45	+42 +43	+44 +44	+45 +46	+43 +43	+45 +45	+46 +46	+43 +42	+44 +44	+46 +45
0.2	+37 +38	+38 +40	+40 +42	+36 +38	+38 +39	+39 +41	+38 +38	+39 +40	+40 +41	+39 +39	+40 +41	+42 +42	+38 +38	+40 +40	+41 +41
0.4	+24 +25	+26 +28	+28 +29	+24 +25	+26 +27	+27 +29	+25 +26	+27 +28	+29 +29	+26 +26	+28 +28	+30 +30	+25 +26	+27 +28	+29 +29
0.6	+15 +16	+18 +19	+19 +21	+15 +15	+17 +18	+19 +20	+15 +16	+18 +19	+20 +21	+16 +17	+19 +20	+21 +22	+16 +17	+19 +19	+21 +21
0.8	+08 +09	+10 +12	+13 +14	+07 +08	+10 +11	+12 +13	+08 +09	+11 +12	+13 +14	+09 +10	+12 +13	+14 +15	+08 +09	+11 +12	+14 +14
1.0	+01 +02	+04 +05	+07 +08	00 +01	+04 +05	+07 +08	+01 +02	+05 +06	+07 +08	+02 +03	+06 +07	+09 +09	+01 +02	+05 +06	+08 +08
1.2	−06 −04	−01 00	+02 +03	−06 −05	−02 −01	+01 +02	−05 −04	−01 00	+02 +03	−04 −04	00 +01	+03 +04	−05 −05	−01 00	+03 +03
1.4	−12 −10	−07 −05	−03 −02	−12 −11	−07 −06	−04 −02	−11 −10	−06 −05	−03 −02	−10 −11	−05 −05	−02 −01	−12 −12	−06 −06	−02 −02
1.6	−17 −16	−12 −10	−08 −06	−18 −17	−12 −11	−08 −07	−17 −16	−11 −10	−07 −06	−16 −17	−10 −10	−06 −06	−18 −18	−12 −12	−07 −07
1.8	−23 −21	−16 −15	−12 −11	−24 −23	−17 −16	−13 −11	−23 −22	−16 −15	−11 −11	−22 −23	−15 −16	−10 −10	−24 −24	−17 −17	−12 −12
2.0	−29 −27	−21 −20	−16 −15	−30 −28	−22 −21	−17 −16	−29 −28	−21 −20	−15 −15	−28 −29	−20 −21	−15 −15	−30 −29	−22 −22	−16 −16

In the simplified theory we assume the following approximations:
1. Sun is a point source at infinity.
2. Attenuation by refraction is small.
3. Terrestrial atmosphere has an exponential structure.
4. Curvature of the light path can be neglected in the computation of the air mass.

The density of the shadow will be given by

$$D' = AM + \log \Phi$$

with [Eq. (2.19)]

$$\Phi = 1 - (a+h) \cos \psi \, \frac{d\omega}{dh_0}, \qquad \omega \approx 0. \tag{2.20}$$

According to the approximations 3 and 4 we have with Eq. (4.10)

$$\omega = c\, \rho(h_0) \sqrt{\frac{2\pi a}{H}}, \qquad M = \frac{H}{c} \omega, \qquad \frac{d\omega}{dh_0} = -\frac{\omega}{H}$$

and the density at the angular distance from the axis

$$r = \psi - \omega$$

will be

$$D' = \frac{AH}{c} \omega + 0.434 \frac{(a+h) \cos \psi}{H} \omega. \tag{2.21a}$$

As a given satellite moves in a limited height interval $h_m \pm \Delta h$ we may write

$$D' = \left[A \frac{H}{c} + \frac{0.434}{H} \sqrt{2 a h_m} \right] \omega + \frac{0.434}{H} \sqrt{2 a h_m} \, \frac{\Delta h - h_0}{2 h_m} \tag{2.21b}$$

where the last term is a correction term.

In the simplified theory the density of the shadow is then a simple function of $\omega = \psi - r$. We can expect that it will be so in an analogous manner in the exact theory where r is replaced by γ and ψ by ψ_0.

$$\sin \psi_0 = \frac{a_\varphi}{a+h}. \tag{2.22}$$

The density should be nearly independent of h and similarly of H i.e. of aerological situation since the influences of its variations are opposite in the first and the second term of Eq. (2.21 b).

Our expectations were fulfilled on these two points and the tables of shadow density for Echo-2 are given here. For it we chose a convenient set of parameters contained in the following survey.

To use these tables, we compute first the angle ψ_0 which we need for the difference $\gamma - \psi_0$ where the angular distance γ is to be obtained from the orbital elements. Then we choose the table having the height (900, 1000, 1100, 1200 km) nearest to the actual height of the satellite during the eclipse. Very rarely an interpolation between two tables may be necessary. In the chosen table we take the situation corresponding most closely to our eclipse. Finally we read in the column under the situation heading the differences $\gamma - \psi_0$ for the given densities D.

2.7. Eclipse Observed from the Satellite

The eclipsed Sun observed from the satellites exhibits approximately similar features as during the sunrise or sunset from the Earth's surface. The mathematical side of the problem is expressed by the Eq. (2.16)

$$r = \pi_\odot \left(1 + \frac{h'_0}{a}\right) + \psi - \omega \qquad (2.16)$$

$$\sin \psi = \frac{a + h'_0}{a + h}.$$

Due to the difference in scale' the light element of the solar disk is here a chord AB (Fig. 2.9) instead of an arc as for lunar eclipses. The whole solar image should be built on by the images $A'B'$ of these chords.

We must further realize the differential influence of the extinction. The lower border of the compressed solar image will be much more strongly attenuated and red coloured than the upper border. However, the course of the sunrise or sunset from the satellite will be very fast, only a fraction of a minute in contrast with the surface phenomenon which lasts on average 2–3 min.

As the local structure of the atmosphere intervenes on the imaging action, some differences may be expected between the tropical and polar regions.

In this connection we may consider the geographical circumstances of the eclipse. They are, however, entirely analogous with those of lunar eclipses (1.1.3).

We trace at first the shadow terminator in the manner already described (1.1.3). Then we calculate the latitude of the point V where the Sun is rising or setting as observed from the satellite

$$\sin \varphi_V = \cos \delta_\odot \cos P. \qquad (1.8)$$

The position angle P is to be computed from the spherical triangle $S'AN$ (Fig. 2.10) by

$$\cos P = \frac{\sin \delta_0}{\sin \sigma \cos \delta_\odot} + \frac{\operatorname{tg} \delta_\odot}{\operatorname{tg} \sigma} \qquad (2.23\,\text{a})$$

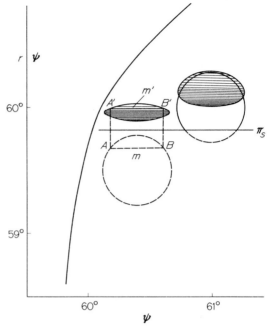

Fig. 2.9. Eclipse of the Sun as observed from the satellite at 1000 km

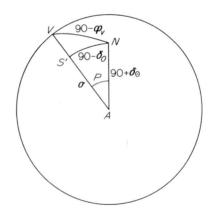

Fig. 2.10. Geographical circumstances of the eclipse

which gives

$$\sin \varphi_V = \frac{\sin \delta_0}{\sin \sigma} + \frac{\sin \delta_\odot}{\operatorname{tg} \sigma}. \qquad (2.23\,\mathrm{b})$$

The angle P is also the angle of position giving the orientation of the vertical solar diameter (i.e. directed towards the Earth's center) in respect to the North.

2.8. Secondary Illuminations

The close proximity of the Earth brings the danger of parasitic light due to the scattering on the surface and in the atmosphere. To estimate the importance of the terrestrial illumination we shall make use of measurements of the Earth albedo. We have then

$$e_\odot = b_\odot \Omega_\odot \quad \text{solar illumination on the satellite}$$
$$e_\mathrm{C} = b_\delta \Omega_\delta \quad \text{terrestrial illumination on the Moon} \qquad (2.24)$$
$$e_\delta = b_\delta \Omega'_\delta \quad \text{terrestrial illumination on the satellite}$$

where Ω are the solid angles and b the luminances of the three bodies. The desired ratio of parasitic to solar illumination will be

$$\frac{e_\delta}{e_\odot} = \frac{b_\delta}{b_\odot} \frac{\Omega'_\delta}{\Omega_\odot}. \qquad (2.25)$$

The unknown ratio of luminances $b_\delta/b_\odot$ may be derived from the albedo measurements which give, for instance, the difference of magnitudes

$$m_\delta - m_\odot = \Delta m \qquad (2.26)$$

as observed from the New Moon. In this situation we have

$$\frac{e_\odot}{e_\mathrm{C}} = 2.512^{\Delta m} \qquad (2.27)$$

which gives together with Eq. (2.25) the ratio

$$\frac{e_\delta}{e_\odot} = 2.512^{-\Delta m} \frac{\Omega'_\delta}{\Omega_\delta}. \qquad (2.28)$$

According to DANJON we have for $\Delta m = 10^m$ at the New Moon i.e. at the Full Earth. At the moment of the eclipse this difference would be much larger at least 14^m and for a satellite at $h = 500$ km we obtain

$$\frac{e_\delta}{e_\odot} < 0.01. \qquad (2.29)$$

The value of Δm has been obtained from Danjon's formula

$$\Delta m = 10^m + 0.022^m \, \varphi \qquad (2.30)$$

where φ is the phase angle. This formula has been directly verified to $-120°$ and we have extrapolated it to $\varphi = 180°$. In fact the Earth's crescent should be extremely narrow as seen from the satellite at the border of the shadow and a fortiori deep in it.

In other words for this relatively low altitude of the satellite and the the Earth's magnitude extrapolated to the New Earth, which is largely overestimated, there is about 1% of the parasitic light of terrestrial origin.

Another source of parasitic illumination lies in the twilight ring. The observer placed on the satellite sees the Earth surrounded by this ring of variable intensity according to the height and the position relative to the Sun. In our estimation we adopt the average value of the luminance $b = 10^{-5} b_\odot$, the radial extension $h = 60$ km on one half of the shadow terminator and the satellite parallax $\pi_s = 60°$. The ratio of terrestrial and solar illumination will then be

$$\frac{e_\delta}{e_\odot} = 5 \cdot 10^{-3}. \qquad (2.31)$$

Again the influence is negligible.

2.9. Fesenkov's Treatment of the Eclipse Problem

Fesenkov's treatment exhibits marked deviations from the purely theoretical form and gives an impression of a more descriptive than exploratory theory [5].

As the attenuation factors during the eclipse he considers three kinds, namely:

a) Attenuation by the differential refraction called here "factor of refractional dispersion" f.
$$(2.32)$$
b) Absorption in ozone layer τ_1.

c) Attenuation by aerosols τ_2.

Nowhere in Fesenkov's paper is there any explicit reference to molecular scattering but it seems that molecular and aerosol scatterings are gathered into one term which is equivalent to the air mass in Rayleigh's atmosphere polluted by aerosols and expressed in zenithal units. To attain a compromise between the pure air mass and the pollution diminishing with altitude, FESENKOV does not choose for the scale height H its aerological value $H = 8$ km but puts in the aerological function

$$\rho = \rho_0 \exp(-h/H) \qquad (2.33)$$

successively $H = 10$, 20 or 40 km in order to select the best fitting for observations, which take place for $H = 20$ km. This polluted air mass M' decreases with the height less rapidly than the pure air mass M and, the greater the value of H adopted, the slower the decrease of $M'(h_0)$ with height h_0.

Secondary Illuminations

If we adopt with FESENKOV the optical density in zenith 0.066 we obtain the grazing densities $D'(h_0)$ (Table 2.3) while in the Rayleigh's

Table 2.3

h_0	10	20	30	40	50
$D(h_0)$	1.43	0.30	0.06	0.01	0.00
$D'(h_0)$	1.87	1.01	0.67	0.38	0.25

atmosphere we have with the same value at zenith 0.066 the densities $D(h_0)$. It is questionable whether the great values of $D'(h_0)$ above 30 km may be admitted even in a normally polluted atmosphere.

No less questionable is Fesenkov's conception of the ozone layer with the distribution function

$$o(h) = h^{-b} \exp(-c h) \qquad (2.34)$$

with $b = 19.21$ and $c = 0.9149$ (h being expressed in km) giving the vertical amount

$$\int_0^\infty o(h) \, dh = 12.08$$

and for the ratio grazing vertical amount the values:

h_0	10	14	16	20	24	30	34 km
	35.57	47.84	53.25	50.58	31.49	7.43	1.93

According to different observed forms of $o(h)$, we usually find in the range 10–20 km an approximate constancy of the above ratio (1.2.17) while Fesenkov's values show there a big rise which seems somewhat unrealistic.

The illumination integral has the usual form

$$e = \int f \exp(-\tau_1) \exp(-\tau_2) \, di \qquad (2.35)$$

with the three above-given attenuation factors and the limb darkening intensity of solar strip di. FESENKOV gives its application to the eclipse of Echo-2 photographically observed at Alma Ata on December 23, 1965, with a 50 cm Maksutov telescope. No indication about the spectral band used can be found with exception of $\lambda = 5000$ Å given in connection with the numerical example of different values of di. Different theoretical curves were computed (Fig. 2.11) to choose the best fitting parameters i.e. $H = 20$ km and optical densities in zenith 0.066 for the polluted atmosphere and 0.022 for ozone. FESENKOV considers his results as

10 Link, Eclipse Phenomena

provisory and claims the necessity for two color measurements inside and outside the Chappuis ozone bands.

The computation of the refraction are similar to those used for lunar eclipses. They are based on CIRA atmosphere but no numerical value of the refraction is given in order to compare it with other results.

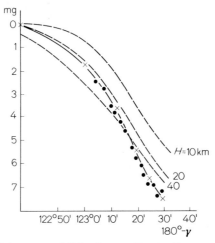

Fig. 2.11. Eclipse light curve of Echo-2 according to FESSENKOV (dots) compared with his theoretical models with different values of H

2.10. Observing Methods Used for Passive Satellites

Up to the present time the photoelectric method has mainly been used for the passive satellites. These satellites are of the category called balloon satellites, e.g. Echo or Pageos whose light intensity is strong enough to allow the use of small photometric apparatus. As an example we shall give here the description of the double photometer used at the Ondrejov Observatory [6].

Each component of the photometer (Fig. 2.12) consists of an object glass O of 50 mm in diameter and of 600 mm in focal length covered by colour filter in a revolving mount F. In the focal plane the diaphragm C limits the field to a circle of $0.75°$ in diameter on the sky. The light from the satellite is then collected by the Fabry lens L which projects the image of the object glass on the cathode FK' of the multiplier EMI.

The photoelectric currents of both photometers can be recorded by two compensating milivoltmeters EZ-2 in order to get simultaneously two light curves in spectral regions transmitted by both color filters. We can, however, obtain both light curves on the same recording chart using the rapid recorder Kipp "Micrograph". The two photoelectric currents are alternately connected by means of a relay system controlled by an astronomical clock to this rapid recorder so that we obtain alternately every two seconds fragments of light curves in two colors exactly timed by the clock.

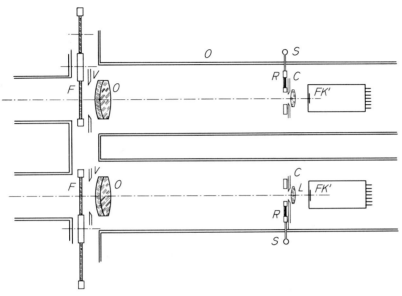

Fig. 2.12. Satellite photoelectric photometer at Ondřejov Observatory

2.11. Work at the Naval Ordnance Test Station, China Lake, California

VENKATESWARAN, MOORE and KRUEGER [7] were the first to have performed the photometry of the eclipses of Echo-1 some months after its launching in August 1960. They used two identical photoelectric photometers for night glow researches with 5-inch refracting telescopes working in two narrow spectral bands centered at 5900 and 5300 Å i.e. in the maximum of Chappuis absorption bands of the ozone and far from it. They obtained on chart records two separate light curves enabling us to determine the differential extinction during the eclipse.

The reduction method used can be described in accordance with our definitions in this book in the following manner. The whole atmosphere is divided into 7 spherical shells with mean heights:

$h_1 = 65$, $h_2 = 55$, $h_3 = 45$, $h_4 = 35$, $h_5 = 25$, $h_6 = 15$, $h_7 = 5$ km

each thus having a 10 km width.

The Sun is assimilated to a point source at its center and the refraction is at first neglected. The partial paths of rays passing for instance, at the grazing height $h_3 = 45$ km, i.e. the mid point of the 3rd shell, are in consequence (Fig. 2.13) in different shells

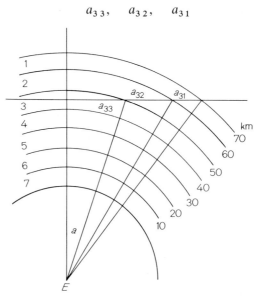

Fig. 2.13. Atmospheric shells according to Venkateswaran

and they may be computed by the simple geometrical formulae

$$a_{33} = 2\sqrt{(a+50)^2 - (a+45)^2}; \quad a_{32} = 2\sqrt{(a+60)^2 - (a+45)^2} - a_{33}$$
$$a_{31} = 2\sqrt{(a+70)^2 - (a+45)^2} - a_{32} - a_{33}.$$
(2.36)

With mean ozone concentrations $x_1, x_2, x_3, \ldots, x_7$ in different shells the optical density due to this gas along the ray $h_3 = 45$ km is expressed in both spectral bands $\lambda = 5900$ Å and $\lambda' = 5300$ Å by

$$B(x_1 a_{31} + x_2 a_{32} + x_3 a_{33}); \quad B'(x_1 a_{31} + x_2 a_{32} + x_3 a_{33}) \quad (2.37)$$

where B and B' are the corresponding absorption coefficients.

In an analogous manner is expressed the extinction due to the molecular scattering

$$A(\rho_1 a_{31} + \rho_2 a_{32} + \rho_3 a_{33}); \quad A'(\rho_1 a_{31} + \rho_2 a_{32} + \rho_3 a_{33}) \quad (2.38)$$

with $\rho_1, \rho_2, \rho_3, \ldots, \rho_7$ air densities in different shells and the extinction coefficients A and A' in both spectral bands. The differential optical density at the satellite position corresponding to the given grazing ray h_3 will then be:

$$D_3 - D'_3 = (A - A')(\rho_1 a_{31} + \rho_2 a_{32} + \rho_3 a_{33})$$
$$+ (B - B')(x_1 a_{31} + x_2 a_{32} + x_3 a_{33}). \quad (2.39)$$

We may in addition allow for the differential absorption along the path between the satellite and the observer which is, however, of minor importance. The left side of the equation is given by the observations

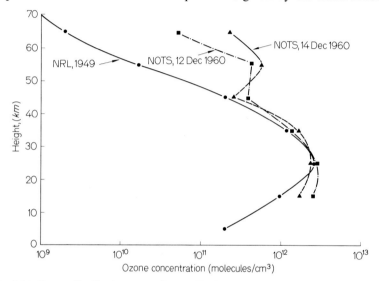

Fig. 2.14. Ozone distribution according to Venkateswaran (NOTS) and the rocket (NRL-1949)

and the first term of the right side can be computed in a given model of Rayleigh's atmosphere. For seven points of the light curve corresponding to seven heights $h_1, h_2, \ldots, h_7$ we have, therefore, a system of seven equations

$$D_1 - D'_1 = (A - A') \rho_1 a_{11} + (B - B') x_1 a_{11}$$
$$D_2 - D'_2 = (A - A')(\rho_1 a_{21} + \rho_2 a_{22}) + (B - B')(x_1 a_{21} + x_2 a_{22})$$
$$D_3 - D'_3 = (A - A')(\rho_1 a_{31} + \rho_2 a_{32} + \rho_3 a_{33})$$
$$ + (B - B')(x_1 a_{31} + x_2 a_{32} + x_3 a_{33}) \quad (2.40)$$
$$\vdots$$
$$D_7 - D'_7 = (A - A')(\rho_1 a_{71} + \rho_2 a_{72} + \cdots \rho_7 a_{77})$$
$$+ (B - B')(x_1 a_{71} + \cdots x_7 a_{77})$$

and their solution yields the seven unknown quantities $x_1, x_2, \ldots, x_7$.

In spite of the approximation adopted by the authors, the final results, i.e. the distribution x of O_3 with the height, is in good agreement with other methods (Fig. 2.14). We may, therefore, hope that a more refined theory of differential absorption will in future give interesting results, especially, if systematic observations of eclipses are continued.

2.12. Work at Ondřejov Observatory on Echo-2 Eclipses

With the photometrical equipment already described (2.10) [6] two series of eclipse curves have been obtained: the first series from September to November 1964 in blue light and the second series during May 1966 in blue and orange light. Their final reduction depends essentially on good orbital elements enabling us to compute the angular distance γ from the antisun and the altitude h of the satellite at every moment of the eclipse. This delicate task has not been accomplished at the present time. Nevertheless we may give here a specimen of the eclipse of May 17, 1966 observed in both colors. [8]

The observed curves (Fig. 2.15) were compared with our theory (1.2.2) valid for the Rayleigh's atmosphere. All computations were performed with the electronic computer Minsk 22.

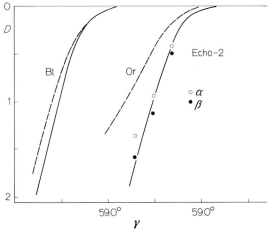

Fig. 2.15. Light curves of Echo-2 eclipses determined at Ondřejov Observatory observed full line, computed dashed line

The curve in blue shows only small differences $O-C$ lying within the limits of possible errors of orbital elements and possibly a small atmospheric pollution would give still better agreement. For the orange curve the differences $O-C$ are more conspicuous. Their origin is to be found in the ozone absorption whose Chappuis bands are located within the transmission region of the orange filter. To take into account the ozone absorption we need to know its distribution with the height $o(h)$ from which the total ozone mass $O(h_0)$ along the ray passing at grazing height h_0 may be computed by simple numerical integration.

We have, therefore, adopted two types α and β of the distribution function $o(h)$ according to HANSEN and MATSUSHIMA [9]. The corresponding total ozone mass is given in the table:

h_0	1	5	10	15	17	19 km
$\alpha\ O(h_0)$	6.23	6.96	8.15	9.34	9.60	9.56 cm
$\beta\ O(h_0)$	8.62	9.64	11.28	12.94	13.29	13.24 cm

h_0	20	22	25	30	40 km
$\alpha\ O(h_0)$	9.40	8.74	7.04	3.71	0.56 cm
$\beta\ O(h_0)$	13.01	12.10	9.75	5.13	0.77 cm

The computed densities Rayleigh's atmosphere + ozone (α) are in rather good agreement with observations at $\gamma = 58.5°$ and $58.7°$, i.e. in atmospheric layers above 25 km. The β-distribution gives there too great a density of the shadow. On the other hand, at $\gamma = 58.3°$, where the layers under 20 km are involved, the observed density is by 0.2 higher than its theoretical value for α-distribution. This difference may be interpreted either by the atmospheric pollution or clouds under 20 km or by the increased amount of O_3 with regard to the α-distribution. For instance, the β-distribution would give a better agreement here, but higher it disagrees clearly with observations. It seems to us, therefore, premature to continue this discussion before the final reduction of all Ondřejov eclipses.

2.13. Work at Valensole Station

RIGAUD [10] investigated the stratospheric absorption between 15 and 30 km by means of the eclipses of Echo-2. From this preliminary account we shall give here some interesting points. The photoelectric photometry was performed in the focus of a Cassegrain telescope using 4 interference filters centred on 3200, 3350, 3440 and 3800 Å i.e. on the beginning of the ozone absorption bands (HUGGINS) and on the atmospheric window where only the Rayleigh's atmosphere and aerosols are acting.

From the light curve of the eclipse we obtain the optical density $D(h_0)$ on the path between the Sun and the satellite which depends on the grazing height h_0 of the ray directed towards the centre of the Sun. The optical density of 1 km path at the height h is computed with the aid of

$$d(h) = -\frac{1}{\pi}\sqrt{\frac{2}{a}} \int_{h_0-h=0}^{+\infty} \frac{\partial D(h_0)}{\partial h_0} d\sqrt{h_0-h}. \qquad (2.41)$$

The curves in Fig. 2.16 give the vertical repartition of $d(h)$ in the spectral band where the ozone absorption should be absent and where, in addition, the effect of the molecular scattering has been removed.

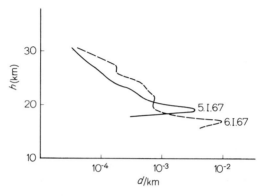

Fig. 2.16. Aerosol extinction as a function of height according to RIGAUD

RIGAUD concludes that there exists an appreciable aerosol pollution in the above height interval, which may be compared with similar phenomena found by ROSEN [11] and VOLZ [12]. Below the maximum at 17—19 km level, the absorption drops rapidly and this may be the origin of a transient brightening of the satellite just before its definitive disappearence known to all observes of satellite eclipses.

2.14. Active Satellites of SR Series

The satellites of Solrad (Solar Radiation) series were developed at the U.S. Naval Research Laboratory [13] with the primary aim of monitoring the invisible U.V. and X-solar radiations especially during flare events. Accessorily these satellites may serve for eclipse purposes.

The main feature of SR-satellites is a 20-inch-diameter aluminium sphere. The sphere is made to rotate around a polar axis determined by the maximum inertial moment about this axis which is larger than the inertial moment about any other axis. To reduce wobble, a damping mercury ring was used.

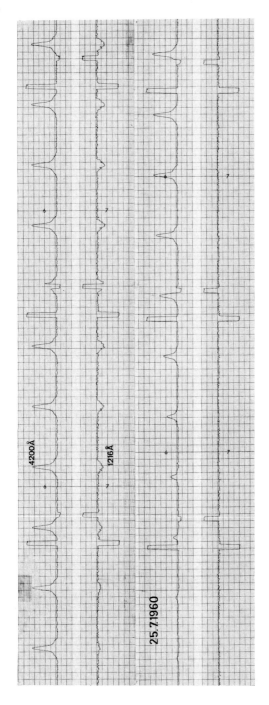

Fig. 2.17. Record of the SR satellite eclipse. Above: *L*-alpha (1216 Å) eclipse, with constant 4200 Å intensity. Below: 4200 Å eclipse

The detectors of different radiations are mounted along the equatorial plane and the spin of the satellite modulates these fluxes. The measured data were telemetered by the usual technique to the NSAA Minitrack network to be recorded there on magnetic tape. Finally at the N. R. Laboratory they were played back so as to obtain the ordinary strip chart (Fig. 2.17).

For the monitoring work the attitude problem of the satellite is very important. In order to determine the aspect angle, i.e. the solar angle above the equatorial plane of the satellite body, one of detectors was a visible light photocell whose indications serve to compute the desired aspect angle. However, during the eclipse this angle is nearly constant and the indications of the aspect detector (for SR-I) may be used for tracing the light curve in visible (violet) light in addition to those in the U.V. or X band.

For the invisible radiations Geiger-Muller counters were used. Filled with different gases and provided with appropriate windows, these ion chambers register different bands of invisible radiations. For X-rays, magnetic shields of Alnico magnets were employed against Van Allen belt electrons.

2.15. Eclipses of the Satellite SR-I

We shall give an account of the observations and of the discussion regarding the eclipses of the SR-I satellite during 1960 [14]. The observational material kindly sent by KREPLIN from the U.S. Naval Research Laboratory at Washington includes:

    i) Light curves in L-alpha light (1216 Å).

    ii) Light curves in violet light (4200 Å).

    iii) Geographical positions and heights of the satellite for several moments during the eclipse.

From the data iii) we first calculated the angular distances of the satellite from the antisolar point and consequently the grazing heights of solar rays going from the center of solar disk to the satellite. The given light curves were then re-drawn with grazing height scale instead of time axis and their principal characteristics are set out in Fig. 2.18.

There is given the distribution in altitude of the following points on every light curve:

    1   the beginning of the decline,
    P   the mid point of the threshold,
    0.5 the half intensity point,
    0   the end of the eclipse.

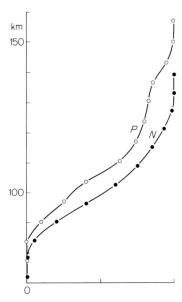

Fig. 2.18. Two L-alpha curves, P perturbated and N normal curve

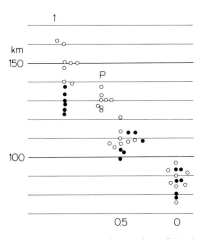

Fig. 2.19. Statistics of L-alpha curves, 1 the beginning of the decline, P the threshold, 0.5 the half-intensity point, 0 the end of the eclipse

The circles are for the perturbated and the dots for normal curves.

One half of all curves (Fig. 2.19) i.e. 7 of L-alpha curves show a threshold, in 2 cases its existence is doubtful and 5 curves are without any perturbation of this kind. To explain the existence of thresholds we may propose the following hypotheses:

a) A relatively regular sequence of instrumental perturbations, regular with regard to the Sun's position over the Earth's limb and limited in time between July 25 and September 2 during some 500 revolutions.

b) Discrete sources of L-alpha radiation located on the solar disk in such a manner as to produce the systematic perturbation of normal curves.

c) Some anomalous transition between the lower region of molecular oxygen and the higher region of atomic oxygen, the former region being responsible for L-alpha absorption. This anomaly could exist between the latitudes 27° S and 73° N in the above time interval.

d) The presence of aerosols in the region between 100–150 km.

The first hypothesis seems very improbable. As to the second hypothesis, the distribution of possible L-alpha sources, i.e. of Ca-plages on the rotating solar disk, must follow a very particular law in connection with the satellite revolution so as to give the perturbations at approximately the same height over the Earth. This would also be an improbable hypothesis.

We must further note that the decline of perturbated curves begins 20 km higher in median values than on normal curves and the height of the threshold is relatively constant between 122–137 km. For these reasons the origin of the thresholds seems to be atmospheric rather than solar.

The L-alpha absorption depends on the distribution of oxygen molecules. We have computed its value with aid of the following formula

$$d(O_2) = 2k \int f(O_2)\, dx, \qquad x = \sqrt{(a+h)^2 - (a+h_0)^2} \qquad (2.42)$$

where the distribution function of O_2 was adopted according to KALMAN-BIJL [15] and the absorption coefficient $k = 0.122 \text{ cm}^{-1}$ (base 10) according to WATANABE [16].

For normal curves we have found a good agreement, but for perturbated ones a drastic modification of the $O_2(h)$ function would be necessary in order to lift the beginning of computed curves by 20 km and in addition the create there the threshold on it. There remains, finally, the hypothesis of an aerosol layer located at the level of about 130 km, or possibly of aerosol clouds. The presence of meteoritic aerosols in the above-mentioned time-interval coincides with the period of higher meteor activity and the temporary elevation of the shadow increase (Fig. 1.5.4).

On the other hand, the violet curves show a regular shape with the decline beginning at about 50 km. The comparison with the theory based on the Rayleigh's atmosphere indicates only a weak pollution of atmospheric layers below 50 km (Fig. 2.20).

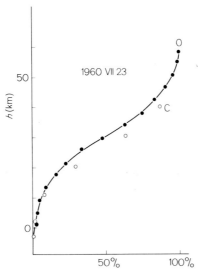

Fig. 2.20. Observed (dots) and computed (circles) light curve in violet light

2.16. Work of Arcetri Group [17]

LANDINI, RUSSO and TAGLIAFERRI have performed at Arcetri Observatory interesting work in eclipse photometry of active satellites belonging to the series of Solrad satellites. The following satellites have been telemetered at ARCETRI.

Designation	
SR-IV 1964:01-D	SR-VII 1965:16-D

Up to now two sets of data have been analysed and used in the study of the upper atmosphere. They are given in Table 2.4.

The starting point of theoretical investigations is, of course, the light curve with the height scale calculated with aid of an ephemeris. This height is defined here in a way common to these problems i.e. the grazing height of the ray going from the satellite to the center of the solar disk, its angular dimension and the refraction being neglected. The calculation of the height and the accuracy of assumed coordinates were checked by comparison with the observed sunset or sunrise in the infrared region by aspect angle detector.

For any point of the light curve we have the equation

$$e = E \exp - \int_{-\infty}^{+\infty} \rho \sigma \, ds$$

Table 2.4

Revolution No.	Date	Event	Satellite
445	February 12, 1964	Sunset	NRL 1964-01-D
1811	May 20, 1964	Sunrise	NRL 1964-01-D
1825	May 21, 1964	Sunrise	NRL 1964-01-D
1839	May 22, 1964	Sunrise	NRL 1964-01-D
1852	May 23, 1964	Sunrise	NRL 1964-01-D
1866	May 24, 1964	Sunrise	NRL 1964-01-D
546	April 18, 1965	Sunset	NRL 1965-16-D
560	April 19, 1965	Sunset	NRL 1965-16-D
601	April 21, 1965	Sunset	NRL 1965-16-D
684	April 27, 1965	Sunset	NRL 1965-16-D
726	April 30, 1965	Sunset	NRL 1965-16-D
740	May 1, 1965	Sunset	NRL 1965-16-D
1686	July 8, 1965	Sunrise	NRL 1965-16-D

where ρ is the number density, an unknown function of the height only, and σ the cross section of the absorbing gas. The above integral can be transformed into Abel's integral equation

$$\rho \sigma = \frac{1}{2\pi} \frac{d}{d\mu} \int^{\mu} \frac{d\left(-\log \frac{e}{E}\right)}{dz} \sqrt{z-u} \, dz \qquad (2.43)$$

with the integration variable

$$z = (a+h)$$

and the integration limits

$$\mu_0 = (a+h_0)^2$$
$$\mu = (a+h_i)^2 \qquad (2.44)$$

where h_0 the grazing height of rays is at the moment when the decline of the light curve just begins and h_i is the height at which we calculate $\rho \sigma$ or ρ. The value of $\log e/E$ can be obtained from the light curve and with the known relation between the time and grazing height h_0 we get

$$\frac{d \log e/E}{dz}.$$

From [Eq. (2.43)] ρ is easily calculated once σ is known from laboratory experiments as the function of the wave length and the molecule involved in the absorption process. In the interval 44–60 Å observed at Arcetri, the value of the cross section σ is nearly independent of the ratio N_2/O, the principal components of the layers concerned, since the cross sections of both particles in the above spectral interval are the same, 2×10^{-19} cm^2, according to VAN ZANDT. Assuming the spectral distribution of grey body at $T = 0.5 \times 10^6$ °K, the effective value of σ can be calculated and consequently the number density ρ.

In this manner the graph of ρ (Fig. 2.21) was obtained using the eclipses of the SR-IV satellite. For the SR-VII satellite the authors have preferred to give only the product $\sigma \rho$ in order to eliminate the influence of a longer observation period and possible different physical conditions in the upper atmosphere (Fig. 2.22).

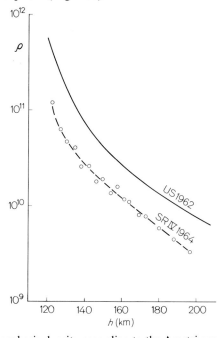

Fig. 2.21. The atmospheric density according to the Arcetri group from 44–60 Å

In the shorter wave length interval 8–14 Å the approximations used for σ are not allowed. However, the value of ρ can be calculated in the case when both curves in 44–60 Å and 8–14 Å intervals were simultaneously observed. Due to the different absorptions in these bands, the light curves cover only a partial interval of altitudes. In this common interval the Eq. (2.43) gives the values of σ and σ' for both spectral bands and consequently the average ratio σ/σ' and finally the values of ρ for the entire altitude range can be obtained.

The following errors which affect the results can be estimated:

15 km in height due to the angular dimensions of the Sun,

5 km in height caused by the inaccuracy of orbital elements,

25% in $\sigma\rho$ at 150 km and 50% at 210 km, due to the uncertainty of the tracings.

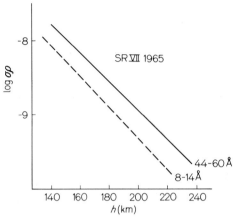

Fig. 2.22. Mean values of $\rho\sigma$ according to the Arcetri group

2.17. Work of Slough Group [18]

THOMAS, VENABLES and WILLIAMS working at the Radio and Space Research Station, Slough, carried out numerous measurements of SR-IV or 1964: 01-D satellite based on the telemetry records at WINKELFIELD during the period January to August 1964. Besides X-measurements during the day in the 0–8, 8–20 and 44–60 Å bands, the authors have obtained some eclipse curves in the last spectral band. This absorption study has been used to test theoretical models of the atmospheric structure in the range 120–200 km.

Their theory may be set out as follows. The basic formula for the optical density in the X-ray region is derived from that for air mass in a thin planetary atmosphere of exponential structure [Eq. (4.9)]

$$\log \frac{E}{e} = 2 N_0 \sigma H_0 \sqrt{\frac{\pi(a+h_0)}{2 H_0}} = D \qquad (2.45)$$

where N_0 represents the particle density and H_0 the scale height at the grazing height h_0. The atmosphere is supposed to have the homogeneous composition of gas having the absorption cross section σ. It is further assumed that the Sun may be assimilated to the point source at its center.

Let us first discuss the influence of the variability of the scale height H which undoubtedly takes place in the range 120–200 km. If we assume with Nicolet [22] the linear variation of H

$$H = H_0 + \beta(h - h_0) \tag{2.46}$$

the particle density of the gas can be expressed by

$$N = N_0 \left[\frac{H_0}{H_0 + \beta(h - h_0)} \right]^{1/\beta} \tag{2.47}$$

and the total gas mass along the grazing ray passing at h_0 is, therefore,

$$2 \int_0^\infty N\, ds = 2 \int_0^\infty N_0\, H_0^{1/\beta}\, \frac{dh}{[H_0 + \beta(h - h_0)]^{1/\beta} \sqrt{(a+h)^2 - (a+h_0)^2}}. \tag{2.48}$$

As we know from previous investigations (1.2.13) the main contribution to the air mass integral comes from the region near the vertex of the light trajectory where $h = h_0$. The slowly varying term $\dfrac{a+h}{\sqrt{2a+h+h_0}}$ may be replaced in the first approximation by $\sqrt{\dfrac{a+h_0}{2}}$

$$2 \int_0^\infty N\, ds = N_0 \sqrt{2(a+h_0)} \int_{a+h_0}^\infty \frac{dh}{\left[1 + \dfrac{\beta}{H_0}(h - h_0)\right]^{1/\beta} \sqrt{h - h_0}} \tag{2.49}$$

and with the substitution

$$\beta/H_0 (h - h_0) = z \tag{2.50}$$

we get the integral

$$2 \int_0^\infty N\, ds = N_0 \sqrt{\frac{2(a+h_0)}{\beta}} \int_0^\infty \frac{dz}{\sqrt{z}(1+z)^{1/\beta}} \tag{2.51}$$

of Euler's kind whose solution gives

$$2\sigma \int_0^\infty N\, ds = 2 N_0\, \sigma \sqrt{(a+h_0) H}\, \alpha(\beta) \tag{2.52}$$

with

$$\alpha(\beta) = \sqrt{\frac{\pi}{2\beta}}\, \Gamma\!\left(\frac{1}{\beta} - \frac{1}{2}\right) \Big/ \Gamma\!\left(\frac{1}{\beta}\right).$$

From (2.52) we obtain for an isothermal atmosphere $\alpha(\beta) = \sqrt{\pi/2}$

$$\frac{d}{dh_0} \log_n D = \frac{d}{dh_0} \log_n \int_0^\infty 2N\, ds = \frac{1}{a+h_0} - \frac{2-\beta}{2H_0}. \qquad (2.53)$$

Denoting

$$\frac{1}{2(a+h_0)} - \frac{d}{da} \log_n D = B \qquad (2.54)$$

and substituting for H_0 in (2.45) we find

$$\sigma N_0 = \frac{D\sqrt{B}}{\sqrt{2(a+h_0)}\,\alpha(\beta)\sqrt{2-\beta}} \qquad (2.55)$$

or with slowly varying function of β

$$\alpha'(\beta) = \frac{1}{\alpha(\beta)\sqrt{2-\beta}} \qquad (2.56)$$

we get finally

$$\sigma N_0 = D\sqrt{\frac{B}{2(a+h_0)}}\,\alpha'(\beta). \qquad (2.57)$$

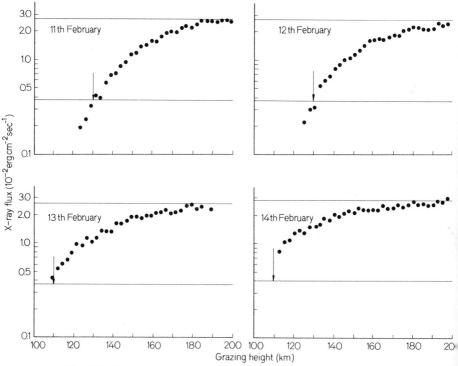

Fig. 2.23. The X-ray flux at 44–60 Å as a function of grazing height

For $\beta=0$, i.e. for the exponential atmosphere, we get the approximation Eq. (2.45). For $\beta=0.2$, we have $\alpha'(\beta)=0.549$ and the error involved in applying Eq. (2.45) is about 3%. Even for $\beta=0.5$ this error amounts only to 10%.

The results obtained during sunsets February 11 – 14, 1964 are shown in the Fig. 2.24 and the corresponding eclipse curves in the Fig. 2.23.

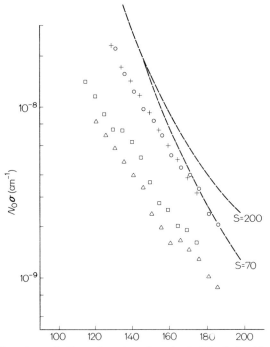

Fig. 2.24. The values of σN_0 as a function of grazing height derived from the measurements in Fig. 2.23. The dashed curves for two solar activities ($S=200$ and 70) computed according to HARRIS and PRIESTER. + 11. II.; ○ 12. II.; □ 13. II.; △ 14. II. 1964

There is a marked difference between the values 11 – 12th and 13 – 14th February, clearly visible in both graphs. This difference may originate from different causes. The first to be discussed is the difference in the position of the vertex of the light paths. While at the first two passes the grazing rays have the geomagnetic latitude of about 40° N, it rises at the following two passes to 53° N. That implies the dependence of σN_0 at a given height h_0 on the geomagnetic latitude with decreasing values towards the pole. When speaking of possible geomagnetic influence, we must consider also, according to the authors, the occurrence of a small geomagnetic storm on 12th February with the maximum on 13th. An explanation of the σN_0 differences in terms of this storm would require that the storm should make a decrease in the atmospheric density in the range of 120–200 km persisting for one day after the geomagnetic disturbance has ended.

Another explanation can be looked for in the distribution of discrete X-ray sources over the solar disk. If we identify these sources with Ca-plages visible on spectroheliograms, their position fails to agree with the behaviour of the curves. An explanation by the non-monochromatic nature of the measurements would require very drastic day-to-day changes in the X-ray spectrum between the above given pairs of days. These changes were not confirmed by the results of X-ray fluxes measured outside the eclipses which were similar on all four days. From our experience we may add that similar day-to-day variations of unknown origin have been found in L-alpha measurements (2.15).

The differences between the observed values and the theoretical ones according the the model given by HARRISON and PRIESTER [21] are probably due to the uncertainties of the parameters involved, especially of the solar activity.

Table 2.5

Date (1964)	Latitude (°N)	Longitude (°E)	$(\sigma N)_{180}$ (10^{-9} cm^{-1})	H (km)
February, 11	34	−46	2.4	25
February, 12	41	−26	2.5	25
February, 13	48	−32	1.1	28
February, 14	51	−37	0.9	28
February, 8	60	20	3.8	22
June, 30	65	15	0.9	22
HARRIS and PRIESTER, 0600 LMT		$S=200$	4.4	19
		$S=70$	2.4	14
HARRIS and PRIESTER, 1800 LMT		$S=200$	4.4	20
		$S=70$	2.7	16

THOMAS and NORTON [19] have further used the eclipse curves to examine the day-to-day changes in density by a much simpler method. They take [19] the X-ray data from six tracking stations of the U.S. Naval Research Laboratory obtained by the SR-IV satellite during February and March 1964. On the eclipse curves the point of the attenuation by a factor of $\exp-1$ was determined and the corresponding grazing height h_g, called the reference grazing height, serves as an indication of an approximately constant pressure level. Its variations will give in consequence an estimation of changes in the atmospheric density.

Fig. 2.25 shows the reference grazing heights obtained from sunset eclipses at three tracking stations, College, Grand Forks and Winkfield,

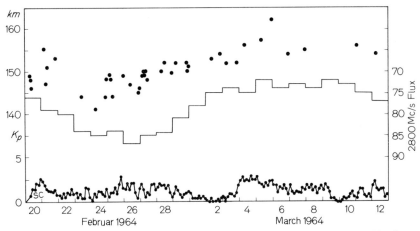

Fig. 2.25. Variations of the reference grazing height (dots), solar 2800 MHz flux (steps) and planetary magnetic index K_p (below)

Table 2.6

Group	Pass No.	Date (1965)	LMT	Latitude (°N)	Longitude (°E)	$(\sigma N)_{180}$ $(10^{-9}\,\mathrm{cm}^{-1})$	H (km)
A	505	April, 15	0456	58	47	7.8	28
LMT 0430	519	April, 16	0441	62	41	3.0	32
Latitude 64° N	533	April, 17	0425	66	35	3.5	33
Longitude 42° E	560*	April, 19	0340	71	45	6.7	25
B							
LMT 0430	520	April, 16	0440	63	15	4.6	30
Latitude 66° N	534	April, 17	0423	66	8	3.6	38
Longitude 17° E	547	April, 18	0402	69	27	4.0	17
C							
LMT 0230	588	April, 21	0242	75	26	7.2	28
Latitude 76° N	602	April, 22	0220	76	18	6.2	23
Longitude 22° E							
D							
LMT 0045	616	April, 23	0133	76	4	5.5	31
Latitude 74° N	643*	April, 24*	0032	77	10	6.2	28
Longitude 2° E	713*	April, 30	0024	73	−4	6.7	34
	727	May, 1	0025	72	−6	5.1	44
E							
LMT 0015	684	April, 27*	0017	76	25	6.9	33
Latitude 74° N	698	April, 28*	0023	75	24	5.3	45
Longitude 28° E	712*	April, 29*	0024	73	22	6.0	36
	753	May, 2*	0022	70	40	5.2	36

over a period of about 20 days. An increase of about 10 km in the grazing height h_g is visible during the period from 23rd February to 5th March. This gives a change of about 60% in density since the scale height near 140 km level is approximately equal to 20 km. In addition there seems

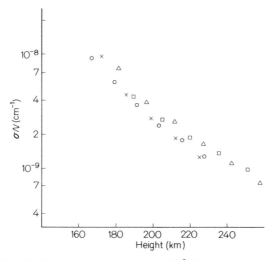

Fig. 2.26. Optical density from 44–60 Å flux in the group D

to exist some relationship between the grazing height and the solar flux at 2800 MHz (Fig. 2.25) but no relation is visible with planetary geomagnetic indices K_p.

VENABLES reports [20] on subsequent measurements obtained by SR-VII satellite 1965-16D in the period from April 15 to May 5, 1965. Seventeen eclipses observed in the 44–60 Å band during this period have yielded the values of σN which were divided into five groups on the basis of date, local time and geographical position of the grazing region (Table 2.6).

It was found that within the uncertainties of the data the results may be represented by the exponential formula

$$(\sigma N)_h = (\sigma N)_{180} \exp \frac{180-h}{H} \qquad (2.58)$$

whose parameters $(\sigma N)_{180}$ and H are listed too, in the above Table 2.6.

In the groups B–D a fairly close agreement between the height variations of σN obtained from different eclipses was found (Fig. 2.26).

The only exception is in the first group A (Fig. 2.27) where the difference between the eclipses No. 505 and 519 is too great to be explained by the uncertainty of the measurements. No substantial difference on the spectroheliograms, as indirect evidence of the distribution and change

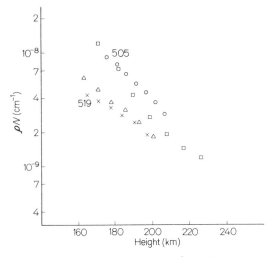

Fig. 2.27. Optical density from 44–60 Å flux in the group A

of X-ray sources, has been observed and Venables ascribes the difference between both eclipses, at least in part, to atmospheric changes.

On the other hand, a comparison of the 1965 results with those of 1964 (Table 2.5–2.6) makes evident three points:

i) In 1965 the density of the neutral atmosphere $(\sigma N)_{180}$ was about twice that in 1964.

ii) Some of the 1964 results of $(\sigma N)_{180}$ could be reconciled with the HARRIS-PRIESTER [21] atmospheric model for the sunspot number $S=70$, but H values lie closer to the model with $S=200$ i.e. for the solar maximum $S=200$.

iii) In 1965 the observations of $(\sigma N)_{180}$ fit the $S=200$ model but for H the agreement is not satisfactory.

In conclusion, the atmospheric density index σN at 180 km changed appreciably during one year around the sunspot minimum epoch 1964/65. The Harris-Priester atmospheric model for sunspot maximum $S=200$ agrees better with the measured values round the solar minimum than does the due minimum model for $S=70$.

Bibliography

1. LINK, F.: BAC **16**, 115 (1965).
2. ŽONGOLOVIČ, I.D., i V.M. AMELIN: Sbornik Tablic Izd. A. N. Moskva 1960.
3. LINK, F.: Space Research **2**, 70 (1962); — BAC **13**, 1 (1962).
4. LINK, F., and L. NEUŽIL: BAC **18**, 359 (1967).
5. FESENKOV, V.G.: Astr. J. S.S.S.R. **44** 3 (1967).
6. LINK, F., and I. ZACHAROV: BAC **17**, 151 (1966).
7. VENKATESVARAN, S.V., J.G. MOORE, and A.J. KRUEGER: J. Geophys. Research **66**, 1751 (1961).
8. LINK, F., L. NEUŽIL, and I. ZACHAROV: Space Research **8**, 86 (1968).
9. HANSEN, J.E., and S. MATSUSHIMA: J. Geophys. Research **71**, 1073 (1966).
10. RIGAUD, P.: Compt. rend. B **265**, 1504 (1967).
11. ROSEN, J.M.: J. Geophys. Research **69**, 4673 (1964).
12. VOLZ, F.E.: Science **14**, 1121 (1964).
13. KREPLIN, R.W., T.A. CHUBB, and H. FRIEDMAN: J. Geophys. Research **67**, 2231 (1962).
14. LINK, F.: Space Research **8** (2), 1220 (1967).
15. KALMAN-BIJL, H.K., and W.L. SIBLEY: Space Research **4**, 279 (1964).
16. WATANABE, K.: Advances in Geophys. **5**, 184 (1958).
17. LANDINI, M., D. RUSSO, and G.L. TAGLIAFERRI: Nature **206**, 173 (1966); — Icarus **6**, 236 (1967) and personal communication.
18. THOMAS, L., F.H. VENABLES, and K.M. WILLIAMS: Plan. Space Res. **13**, 807 (1965).
19. THOMAS, L., and R.B. NORTON: Space Research **7**, 1183 (1967).
20. VENABLES, F.H.: Plan. Space Res. **15**, 681 (1967).
21. HARRIS, I., and W. PRIESTER: J. Geophys. Research **67**, 4585 (1962).
22. NICOLET, M.: J. Atm. Terr. Phys. **1**, 142 (1951).

3. Twilight Phenomena

3.1. Different Components of the Twilight

Although the complete investigation of twilight phenomena involves meteorological optics and aeronomy, there remains nevertheless a certain relationship with lunar eclipses. Like them, the twilight phenomena are in fact due to the Earth's shadow thrown, into space, however, not on the lunar surface but in the terrestrial atmosphere. In other words, if during the lunar eclipses the variable structure of the Earth's atmosphere determines the general features of the eclipse, during twilight its role is double, firstly, as in lunar eclipses, as an attenuating factor and secondly as a scattering agent extended along the line of sight.

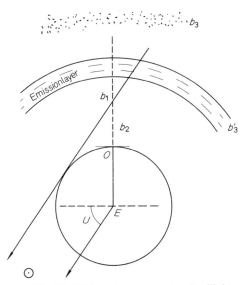

Fig. 3.1. Different components of twilight

The simplest case of twilight phenomena is the luminance of the sky in zenith which is the function of the solar depression angle U (Fig. 3.1). At a given moment the observed luminance is the sum of the following components:

1. *High component* on the part of the vertical sight line directly illuminated by the Sun.

The greatest part of it is the primary scattered light on the air molecules or other particles directly illuminated by the Sun. A minor part is due to the illumination by the sky and the Earth's surface.

2. *Low component* on the part of the sight line immersed in the shadow.

The scattering molecules and particles on this part are illuminated by the sky and the Earth's surface.

The first component is most interesting as it originates in the upper atmosphere. The second component, situated mainly in the lower stratosphere and in the troposphere, may be considered as a perturbing factor whose influence increases with increasing solar depression.

3. *Night sky light* of stellar and atmospheric, i.e. emission, origin enters in the observed luminance at the end of the twilight and becomes absolutely preponderant from the beginning and during the night. The emission or the airglow component may be avoided by the appropriate choice of the spectral band.

3.2. Night Light Correction

The light of the night sky (component 3) — no matter what its origin — adds itself to both scattering components (1 and 2). Its influence is negligible at the beginning and mid-twilight but becomes predominant at the end i.e. when night succeeds to twilight. For the analysis of the twilight phenomena, it is important correct the measurements of this component. Calling b_1, b_2, b_3 the three above components we have an obvious relation for the measured luminance of the twilight sky

$$b = b_1 + b_2 + b_3. \qquad (3.1)$$

It remains to subtract b_3 from the measured luminance b in order to obtain the pure scattering luminance $b_1 + b_2$. As at the end of the twilight $b_1 + b_2$ is nearly equal to zero, we can put the luminance of the night sky equal to the necessary correction with two assumptions:

 i) The airglow is negligible or constant during the whole twilight. That is almost time outside the strong emission bands 0.56, 0.59 and 0.63 µ.

 ii) The stellar background is constant which involves some distance from the Milky Way of the measured place.

To facilitate the night light correction of the luminance values generally expressed in logarithms, we may use the following table [1] which enables us to read directly the correction of the observed value $\log b$.

Table 3.1

Δ	K	Δ	K	Δ	K	Δ	K
0.21	0.416	0.61	0.122	1.02	0.044	1.82	0.007
0.22	0.401	0.62	0.119	1.04	0.042	1.84	0.006
0.23	0.386	0.63	0.116	1.06	0.040	1.86	0.006
0.24	0.372	0.64	0.113	1.08	0.038	1.88	0.006
0.25	0.359	0.65	0.110	1.10	0.036	1.90	0.006
0.26	0.346	0.66	0.107	1.12	0.034	1.92	0.005
0.27	0.334	0.67	0.104	1.14	0.033	1.94	0.005
0.28	0.323	0.68	0.102	1.16	0.031	1.96	0.005
0.29	0.312	0.69	0.099	1.18	0.030	1.98	0.005
0.30	0.302	0.70	0.097	1.20	0.028	2.00	0.004
0.31	0.292	0.71	0.094	1.22	0.027	2.05	0.004
0.32	0.283	0.72	0.092	1.24	0.026	2.10	0.003
0.33	0.274	0.73	0.089	1.26	0.025	2.15	0.003
0.34	0.265	0.74	0.087	1.28	0.023	2.20	0.003
0.35	0.257	0.75	0.085	1.30	0.022	2.25	0.002
0.36	0.249	0.76	0.083	1.32	0.021	2.30	0.002
0.37	0.242	0.77	0.081	1.34	0.020	2.35	0.002
0.38	0.234	0.78	0.079	1.36	0.019	2.40	0.002
0.39	0.227	0.79	0.077	1.38	0.018	2.45	0.002
0.40	0.220	0.80	0.075	1.40	0.018	2.50	0.001
0.41	0.214	0.81	0.073	1.42	0.017	2.55	0.001
0.42	0.208	0.82	0.071	1.44	0.016	2.60	0.001
0.43	0.202	0.83	0.070	1.46	0.015	2.65	0.001
0.44	0.196	0.84	0.068	1.48	0.015	2.70	0.001
0.45	0.190	0.85	0.066	1.50	0.014	2.75	0.001
0.46	0.185	0.86	0.065	1.52	0.013	2.80	0.001
0.47	0.180	0.87	0.063	1.54	0.013	2.85	0.001
0.48	0.175	0.88	0.061	1.56	0.012	2.90	0.001
0.49	0.170	0.89	0.060	1.58	0.012	2.95	0.000
0.50	0.165	0.90	0.058	1.60	0.011	3.00	0.000
0.51	0.161	0.91	0.057	1.62	0.011		
0.52	0.156	0.92	0.056	1.64	0.010		
0.53	0.152	0.93	0.054	1.66	0.010		
0.54	0.148	0.94	0.053	1.68	0.009		
0.55	0.144	0.95	0.052	1.70	0.009		
0.56	0.140	0.96	0.050	1.72	0.008		
0.57	0.136	0.97	0.049	1.74	0.008		
0.58	0.133	0.98	0.048	1.76	0.008		
0.59	0.129	0.99	0.047	1.78	0.007		
0.60	0.126	1.00	0.046	1.80	0.007		

$\Delta = \log$ twilight sky $- \log$ night sky; log corrected twilight sky $= \log$ twilight sky $- K$.

3.3. Illumination of the Upper Atmosphere [2]

Let us compute the solar illumination of a point A in the upper atmosphere at height h and for solar depression U at the foot A' of the vertical line of A (Fig. 3.2). Our task is identical with the problem regarding the eclipses of satellites, the only difference being the much lower heights involved.

In fact, we can use the same expression Eq. (2.13) in order to calculate the illumination at a if we put simply

$$\gamma = 90° - U.$$

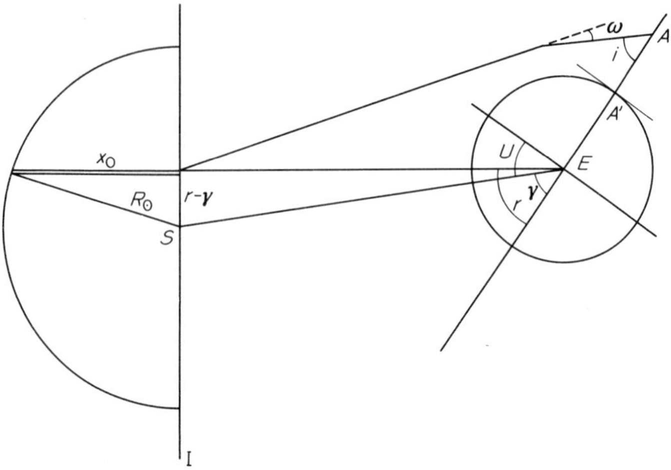

Fig. 3.2. Illumination of the upper atmosphere during twilight

The expression for the illumination will then be as before [Eq. (2.11)]

$$e = 2k \int_{\gamma - R_\odot}^{\gamma + R_\odot} T\left[(1-\kappa)x_0 + \frac{\pi \kappa}{4R_\odot} x_0^2\right] dr. \qquad (3.2)$$

Some authors have neglected the angular dimension of the solar disk putting $e/E = T_0$, i.e. equal to the transmission of the ray issued from the center of solar disk. In order to be more exact, we shall derive the necessary correction depending on the solar disk. To this purpose we write in the above integration interval for the transmission coefficient an approximation

$$T = T_0 e^{-\alpha(r-\gamma)} = T_0\left[1 + \alpha(r-\gamma) + \frac{\alpha^2}{2!}(r-\gamma)^2 + \cdots\right] \qquad (3.3)$$

where the factor α depends on the wave length and the depression U. Putting it in the integral Eq. (3.2) we get

$$T_0 \int_{\gamma - R_\odot}^{\gamma + R_\odot} \left\{ \frac{2(1-\kappa)\alpha^n(r-\gamma)^n \sqrt{R_\odot - (r-\gamma)^2}}{n!} \right.$$
$$\left. + \pi \kappa \alpha^n (r-\gamma)^n [R_\odot^2 - (r-\gamma)^2] \right\} dr$$

Illumination of the Upper Atmosphere

or with the general form of even terms ($n = 0, 2, 4, 6, \ldots$)

$$\pi R_\odot^2 \frac{(3-\kappa)}{3} \frac{R_\odot^2 \alpha^n}{n!} \frac{3}{3-\kappa}$$
$$\cdot \left[2(1-\kappa) \frac{1 \cdot 3 \cdot 5 \ldots (n-1)}{2 \cdot 4 \cdot 6 \ldots (n+2)} + \frac{2\kappa}{(n-1)(n+3)} \right] \quad (3.4)$$

whereas the odd terms ($n = 1, 3, 5, 7, \ldots$) are equal to zero

We have finally for the illumination

$$e = T_0 \left(1 + \frac{R_\odot^2}{40} \frac{15 - 7\kappa}{3 - \kappa} + \frac{R_\odot^4 \alpha^4}{2240} \frac{35 - 19\kappa}{3 - \kappa} + \cdots \right) \quad (3.5)$$

where the illumination outside the atmosphere

$$E = \pi R_\odot^2 \left(1 - \frac{\kappa}{3} \right)$$

is as usually taken as unity.

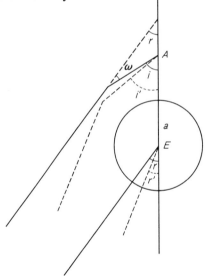

Fig. 3.3. Light paths during twilight

The value of α is computed from two rays both issued from the point A and having their grazing heights h_0 and h'_0. For the first ray we choose the angle r so that $i = r + \omega$ (Fig. 3.3) and the invariance theorem gives the corresponding height of the point A

$$h = h_0 + a[1 - \sin i + c(\rho_0 - \rho)]. \quad (3.6)$$

On the other hand for the second ray we compute the angle i'

$$\sin i' = 1 + c(\rho'_0 - \rho) - \frac{h - h'_0}{a} \qquad (3.7)$$

which gives

$$r' = i' - \omega'. \qquad (3.8)$$

In addition, we determine for both rays the transmission coefficients T and T' so that finally with aid of Eq. (3.3) we obtain

$$\alpha = 2.30 \ldots \frac{\log T' - \log T}{r - r'}. \qquad (3.9)$$

If we express the illumination by the density of the shadow in the upper atmosphere we get

$$D = -\log T_0 - \log\left(1 + \frac{R_\odot^2 \alpha^2}{40} \frac{15 - 7\kappa}{3 - \kappa} \ldots\right) \qquad (3.10)$$

where the last term is the disk correction given in the following table.

Table 3.2

λ (Å)	A	κ	$\alpha=4.5$	3.75	3.0	2.25	1.5
4000	0.0197	0.81	0.30	0.21	0.14	0.08	0.04
5000	0.0179	0.66	0.31	0.22	0.14	0.08	0.04
6000	0.0138	0.55	0.32	0.23	0.15	0.09	0.04
8000	0.0112	0.42	0.33	0.23	0.15	0.09	0.04
10000	0.0105	0.35	0.34	0.23	0.15	0.08	0.04

3.4. Fundamental Problems of Twilight Phenomena

If we consider the twilight phenomena as a tool for the exploration of the upper atmosphere, the fundamental problems of twilight phenomena may be expressed as follows:

i) What is the ratio of the hight component of the twilight luminance to that of the Sun?

ii) What is the ratio of high to low component, the latter being measured in good conditions at the high altitude station (2–4 km)?

Numerous methods elaborated in order to obtain the above ratios are lacking in certainty because of the many hypotheses involved. The only reliable way was started recently by the use of balloon measurements in the stratosphere. We shall give here a brief account of it [3].

The measurements were prepared at Centre d'Aéronomie, Verrières, France, and the balloons were launched from the camp at Aire/Adour near the Pic-du-Midi Observatory (2.9 km) where simultaneous measurements were conducted by a similar method.

The photometer used is built according the normal scheme with the objective covered with an interference filter (5300 Å), the field diaphragm, Fabry lense and the multiplier EMI. The intensity of the photoelectric current was telemetered by the usual method and recorded at the receiving station at Aire. The geographic coordinates of the balloon variable during the flight were determined by the radar service and the height by the meteorological elements. During twilight the intensity variation of $1:10^4$ takes place between night and sunrise. In order to not overload the multiplier and to maintain the deflection of the recoder within the scale, a rotating disk with 10 neutral filters of increasing density (by 0.3) was inserted in the optical path. The change of filters was controlled by the photoelectric current itself controlled by means of a discriminator.

The optical axis of the photometer was inclined by 30° from the zenith and the photometer was rotated round the vertical axis at a speed of 6–8 revolutions p.m. In this way, the photometrical profile at 30° zenithal distance and in different azimuths was recorded from night to sunrise at the balloon ceiling which was in average at 25 km height.

A similar photometer but without automatic device and recorder was used at the Pic-du-Midi Observatory (2.9 km) in order to take simultaneous measurements with balloons. The two apparatuses were edibrated together so as to give the values of twilight luminances expressed in the same scale.

In addition, the Pic-du-Midi photometer was used to measure the light intensity of the Sun during the day. These measurements give Bouguer's straight lines whose extrapolation leads to the extraterrestrial intensity of the Sun. In this way the luminance of the twilight sky at 25 km level was indirectly related to that of the Sun. The direct operation would be more difficult owing to the necessity of guiding the balloon photometer towards the Sun.

During the summer of 1966, 5 balloon ascents were carried out four of which were doubled by the simultaneous measurements at Pic-du-Midi where, in addition, about 15 Bouguer straight lines were determined. In the first place, the results gave the comparison of luminances measured at 25 an 2.9 km levels. The luminances at Pic-du-Midi (Fig. 3.4) were always higher at the balloon level. The main cause of this is the admixture in the total light of the low component, the importance of which rapidly decreases with the height. In the first approximation we may assume the low component to be proportional to the air mass on the line of sight. We have at 30° from the zenith:

Air mass above the balloon at 25 km $M_1 = 0.23$ km
Air mass above Pic-du-Midi at 2.9 km $M_2 = 6.35$ km
Air mass between 25 and 2.9 km $M_2 - M_1 = 6.12$ km

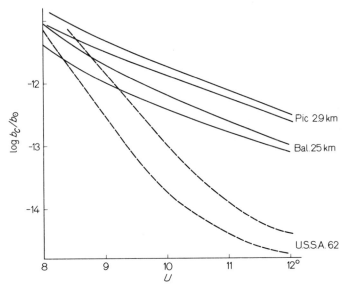

Fig. 3.4. Twilight curves observed at Pic-du-Midi, from balloon, and computed in the U.S. Standard Atmosphere 1962

If the above assumption may be adopted we have for the remaining low component above the balloon level

$$s_1 = \frac{b_2 - b_1}{M_2 - M_1} M_1 = 0.038 (b_2 - b_1) \tag{3.11}$$

where b_2 and b_1 are the luminances measured simultaneously at both levels. To be on the safe side should any error have been committed in the above assumption, we may raise the above factor 0.038 to 0.1 and we get

$$p' = b_1 - s_1 = 1.1 b_1 - 0.1 b_2. \tag{3.12}$$

Having obtained the corresponding corrected values of the high component, we shall compare it with the theoretical values. The formula of RAYLEIGH-CABANNES [4] for molecular scattering gives the ratio of both twilight and solar luminances in the following expression

$$\frac{b_c}{b_\odot} = \frac{12\pi^3}{N} \frac{c^2}{\lambda^4} \frac{1+\delta}{6-7\delta} \left(1 + \frac{1-\delta}{1+\delta} \cos^2 i\right) R_\odot^2 \int 10^{-D} \rho \, ds \quad (3.13)$$

where N the number of air molecules in 1 cm$^3 = 2.69 \times 10^{19}$
    c the refractivity of the air $= 0.000293$
    δ the factor of depolarisation $= 0.042$
    λ the wave length ($= 5300$ Å)
    ρ the air density
    i the scattering angle
    $R_\odot$ the angular radius of the Sun
    D the optical density of the twilight shadow
    ds the linear element of line of sight.

The geometry of the twilight scene enables us to calculate all quantities involved in the above formula at every azimuth A from the Sun and for every solar depression. The density D of the twilight shadow was taken from our tables [2] and for the air density we adopted the values given by the U.S. Standard Atmosphere.

As it appears from Fig. 3.4, theoretical luminances are generally inferior to the observed values. The ratio of both p'/p increases with the increasing altitude of the illuminated part of the upper atmosphere. We can explain this in to ways. The first explanation could be sought in wrong values of the air density extracted from the U.S. Standard Atmosphere Tables. However, the necessary corrections are somewhat too great to be accepted without further confirmation by another method.

The second explanation seeks the origin of higher luminances in the presence of aerosols, probably of meteoritic composition, forming at a level of some 100 km a layer or polluted region as is indicated by lunar eclipses. The light scattering in this layer becomes more and more important with increasing height or with decreasing air density.

This second hypothesis seems to be supported by the variation of the luminance with the azimuth. For the depression interval $11° > U > 9°$ when the layers between 150 and 80 km are directly illuminated the observed amplitude of the azimuth profile (Fig. 3.5) is much smaller than its theoretical value.

Again, according to the first explanation, the corrections of the U.S. Standard Atmosphere would be required not so much for the air density as for the relatively large diminution of its gradient with height, i.e. the large increase of the scale height which does not seem to be acceptable at the levels involved.

On the contrary, the second hypothesis which claims the presence of aerosols offers an adequate explanation without necessitating any

178   Twilight Phenomena

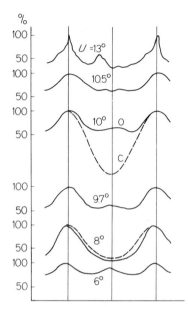

Fig. 3.5. Photometrical profiles of the twilight sky at 30° from zenith in different azimuths, obtained at 25 km

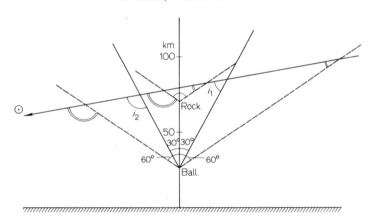

Fig. 3.6. Comparison of balloon and rocket measurements

modification of the above mentioned tables. The secondary maximum, too, (Fig. 3.5) opposite the Sun is unexplained in Rayleigh's atmosphere. This anomalous behaviour of the azimuth profiles is limited to the layers above 80 km. Beginning with about $U=8.5°$ both the theoretical and observed curves are in good agreement (Fig. 3.5). In other words, the location of light excess is clearly limited to the upper atmosphere for $h > 80$ km. Unfortunately enough, the higher layers cannot be explored because of the night sky component whose light becomes prevalent from $U = 13°$.

The amplitude of the azimuth profile depends on two factors, i.e. on the difference of the air density or on its gradient in the range of some 50 km above the limit of the shadow (Fig. 3.6), and on the scattering function of the angle i. In the above experiment it is difficult to separate both influences. One might suggest the choice of a greater zenithal distance of observation in order to increase the difference of scattering angles in both extreme azimuths. Unfortunately this method is not efficient for as a consequence the height difference would rise simultaneously (Fig. 3.6). A more efficient method would be to go higher with the observing point B from 25 km close to the limit of the shadow where the increase of the zenithal distance involves only a slight increase in the height difference in both extreme directions. This is, however, a problem of rocket photometry as no balloon can attain these altitudes.

Bibliography

1. NEUŽIL, L.: Stud. Geoph. Geod. **8**, 325 (1964).
2. LINK, F.: Mitt. Beob. Tschech. Astr. Ges. Prag **6**, 1 (1941).
3. LINK, F., L. NEUŽIL, et I. ZACHAROV: Ann. géophys. **23**, 207 (1967).
4. CABANNES, J.: Sur la diffusion de la lumière. Paris 1929.

4. Occultations and Eclipses by Other Planets

4.1. General Remarks

We shall first give the classification of phenomena dealt with in this chapter.

a) Occultation of a star by the planet observed from the Earth, called the far occultation.

b) The same phenomenon but observed from a spacecraft orbiting or flying near the planet, i.e. the near occultation.

c) Eclipses of satellites or spacecrafts by the shadow of the planet illuminated by the Sun.

There are many analogies between the eclipses of the Moon and the above phenomena. These is also a basic difference in the approach to the theoretical investigations. In lunar eclipses a relatively good knowledge of the atmospheric structure provides us with a solid framework for the exploration of the high layers. On the other hand, in the planetary cases we aim at a basic knowledge of the atmospheric structure, leaving detailed studies to some time in the near future.

There are also some differences in the observational material we have actually at our disposal. In fact we have observed only about half a dozen planetary phenomena in contrast with many tens of terrestrial ones of much higher quality. We must, therefore, hope that the future development of astronautics will provide us with necessary observational data.

4.2. Dioptrics of the Thin Planetary Atmosphere

In far occultations only the highest regions of the planetary atmosphere intervene (4.4) where the atmosphere is relatively thin. This property, in addition to the assumption of the exponential structure, offers some advantages in the computation of the refraction and the air mass.

We have consequently

$$\rho(h) = \rho_0 \exp -\frac{h}{H} \qquad (4.1)$$

Dioptrics of the Thin Planetary Atmosphere

where the scale height is given as ordinary by

$$H = \frac{kT}{mg} \tag{4.2}$$

with k the gas constant
  T the absolute temperature
  m the molecular mass
  g the acceleration of gravity on the planet.

The elements of the refraction and of the air mass are as already derived [Eq. (1.40)]

$$d\omega = \frac{d\mu}{\mu} \operatorname{tg} i \tag{4.3}$$

$$dM = \rho \, ds.$$

We may assume the straight form of the light trajectory (Fig. 4.1) and it follows

$$\operatorname{tg} i = \frac{a + h_0}{s} \tag{4.4}$$

with

$$s = \sqrt{2a(h - h_0)} \tag{4.5}$$

and

$$ds = \frac{1}{2}\sqrt{\frac{2a}{h - h_0}} \, dh. \tag{4.6}$$

The Eq. (4.3) take the form with

$$\mu = 1 + c\rho, \qquad c \approx 3 \times 10^{-4}, \tag{4.7}$$

$$d\omega = c\,\rho(h_0)\sqrt{\frac{2a}{H}} \exp(-x^2) \, dx \tag{4.8}$$

$$dM = \rho(h_0)\sqrt{2aH} \exp(-x^2) \, dx$$

$$x = \sqrt{\frac{h - h_0}{H}}$$

$$\rho(h_0) = \rho_0 \exp{-\frac{h_0}{H}}$$

and their integration in the limits from $-\infty$ to ∞ gives

$$\omega = c\,\rho(h_0)\sqrt{\frac{2\pi a}{H}} = \omega_0 \exp{-\frac{h_0}{H}} = c\,\frac{M}{H}$$
$$M = \rho(h_0)\sqrt{2\pi a H} = M_0 \exp{-\frac{h_0}{H}} = H\,\frac{\omega}{c} \tag{4.9}$$

where ω_0 and M_0 are the values at the level $h = 0$.

In addition we may compute

$$\frac{d\omega}{dh_0} = -\frac{\omega}{H} = -\frac{c}{H^2}M$$

$$\frac{dM}{dh_0} = -\frac{M}{H} = -\frac{\omega}{c}.$$

(4.10)

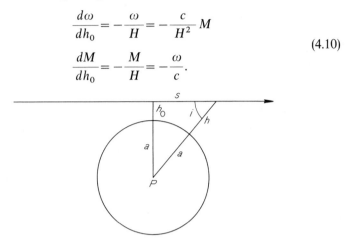

Fig. 4.1. Light trajectory in a thin atmosphere

4.3. Planetary Atmosphere from the Refraction Standpoint

Let us consider the ray leaving the planetary surface under the zenithal distance z (Fig. 4.2). At every point of the trajectory the invariance relation is valid

$$a\,\mu_0 \sin z = r\,\mu \sin i. \tag{4.11}$$

One may ask if every ray can leave the planet [1, 2]. The necessary condition for it is

$$\frac{di}{dr} < 0 \tag{4.12}$$

i.e. the angle i should always decline with increasing distance r.

By deriving Eq. (4.11) we get

$$\cos i\, di = -\frac{a\,\mu_0 \sin z}{\mu^2 r^2}(\mu\, dr + r\, d\mu). \tag{4.13}$$

With the known relation

$$\mu = 1 + c\,\rho \tag{4.14}$$

and the assumed exponential structure in a limited height interval

$$\rho = \rho_0 \exp{-\beta h} \tag{4.15}$$

we obtain

$$\frac{di}{dr} = -\frac{a\,\mu_0 \sin z}{\mu^2 r^2 \cos i}[1 + c\,\rho - (a+h)\,c\,\beta\,\rho]. \tag{4.16}$$

If the expression

$$1 + c\rho - (a+h)c\beta\rho > 0 \tag{4.17}$$

is positive the condition Eq. (4.12) is fulfilled and the ray can leave the atmosphere to go into space. Such is the case of the terrestrial atmosphere.

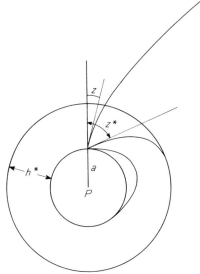

Fig. 4.2. Light trajectories in a dense atmosphere

On the other hand, when

$$1 + c\rho - (a+h)c\beta\rho < 0, \tag{4.18}$$

we must consider also the zenithal distance z as the criterion for the escape of the ray.

From (4.18) we get

$$\rho > \frac{1}{(a+h)c\beta - c} = \rho^* \approx \frac{1}{ac\beta}. \tag{4.19}$$

In this part of the atmosphere the angle i grows as the beginning and reaches its maximum

$$\sin i^* = \frac{\mu_0 a}{\mu^* r^*} \sin z = \sin z \left[1 + c(\rho_0 - \rho^*) - \frac{h^*}{a} \right] \tag{4.20}$$

at the level h^* where the density is ρ^*.

Having passed this level under the angle $i^* < 90°$, the ray goes out into space. In the case when $i^* = 90°$ the ray cannot leave the planetay r atmosphere and returns to the surface. The limiting zenithal distance for it is now

$$\sin z^* = 1 - c(\rho_0 - \rho^*) + \frac{h^*}{a}. \qquad (4.21)$$

Further rays whose $z > z^*$ remain imprisoned in the atmosphere between the surface and the level $h_1 < h^*$. With assumed height of the summit h_1 and the corresponding density ρ_1 we compute the zenithal distance

$$\sin z_1 = 1 - c(\rho_0 - \rho_1) + \frac{h_1}{a}. \qquad (4.22)$$

Thus we have two categories of rays:

i) The rays whose $z < z^*$ which leave the planet

ii) The rays whose $z > z^*$ which cannot leave the planet.

The rays issued under the zenithal distance z^* are approaching asymptotically to the height h^* which we call the critical level. In other words for an occultation or eclipse the atmosphere behaves as if it were opaque up to the critical level h^*. There is some chance of Jupiter's and also Venus's atmosphere having the critical level.

4.4. Basic Equations for the Far Occultation

This already classical case of occultation was investigated by PANNEKOEK [3] and again quite independently by FABRY [4]. As will be seen later on, only the very thin part of the planetary atmosphere is to be considered. We may, therefore, use the formulae developed above (4.2) (Fig. 4.3).

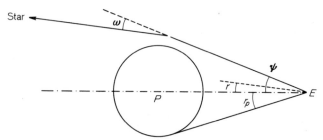

Fig. 4.3. Occultation scene

The attenuation by refraction is obtained from Eq. (1.35) putting there $\omega \to 0$ and with use of Eq. (4.10)

$$\Phi = 1 + \frac{l}{H} \omega \qquad (4.23)$$

where l is the distance of the planet from the Earth.

The geometrical angular distance of the occulted star from the centre of the planet will be according to Eq. (1.11)

$$r = r_p \left(1 + \frac{h_0}{a}\right) - \omega \tag{4.24}$$

where r_p is the angular radius of the planet as seen from the Earth.

The apparent angular distance will then be

$$\psi = r_p \left(1 + \frac{h_0}{a}\right). \tag{4.25}$$

We have neglected the extinction. If we compare the attenuation by refraction [Eq. (4.23)] with the approximate expression for the attenuation by extinction

$$10^{-AM} \approx 1 - \frac{2.30\, AH\omega}{c} \tag{4.26}$$

we see that

$$\frac{l}{H}\omega \gg 2.30\,\frac{AH\omega}{c} \quad \text{with } l \approx 10^8 \text{ km.} \tag{4.27}$$

In other words, the occulted star attenuated by refraction disappears before the extinction can attain a measurable amount. As the distance l is large, the disapperance takes place according to Eq. (4.27) for small values of the air density. The far occultations are useful only for the exploration of the thin parts of planetary atmospheres.

The above given Eq. (4.23)–(4.25) describe completely the course of the occultation when the geometrical distance r is computed from the ephemeris. Then it is advantageous to choose as reference the time and the position of the planet when the attenuation by refraction attains the value $\Phi = 2$. Then the refraction

$$\omega^* = \frac{H}{l} = \mathcal{H}$$

is equal to the angular value of the scale height H as seen from the Earth. Introducing the function $F = \Phi - 1$ we get

$$r = r_p^* - \mathcal{H}(\log F + F)$$
$$\psi = r_p^* - \mathcal{H} \log F \tag{4.28}$$

where the logarithms are natural.

4.5. Theoretical Course of a Far Occultation

We shall give here the theoretical course of the occultation by Mars. We adopted $H = 22$ km and we suppose the central occultation with the planetary motion to be $dr/dt = 0.025''/\text{sec}$. Thus according to Eq. (4.24) we obtain for the geometrical distance

$$r = \psi - \omega = 0.025''(t - t_0) \tag{4.29}$$

where t_0 is the moment of the central occultation when $r = 0$. For the apparent distance Eq. (4.29) yields

$$\psi = 0.025''(t - t_0) + \omega. \tag{4.30}$$

Before the ingress far from the limb ω is small; both the geometrical and the apparent distance are identical, varying uniformly with the time. Nearer to the limb the influence of growing refraction slows down

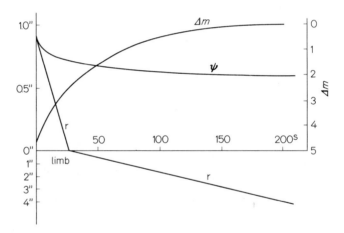

Fig. 4.4. Theoretical course of a far occultation

the descent of the curve of ψ and practically stops it. The occulted star seems to be caught by the planetary limb and carried by the planetary motion. If the occulted star were bright enough we could follow it at the limb when actually it is far behind the limb. In fact the attenuation by the refraction causes the star to disappear from our sight (Fig. 4.4).

Theoretically the two following things may occur, depending on the attenuation of the refraction at the level of complete opacity (clouds or soil)

$$\omega_0 \leq r_p. \tag{4.31}$$

In the first case the star could really disappear behind the limb. In the second case the occulted star could theoretically remain visible even when considerably attenuated throughout the whole occultation. Is image deformed by the refraction could travel round the limb as is shown in Fig. 4.5.

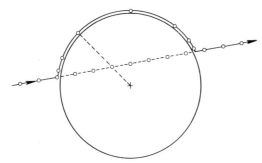

Fig. 4.5. Apparent positions of the star during the non central occultation

4.6. Occultation of Regulus by Venus, July 7, 1959

This very rare phenomenon — occurring only four times between 600 AD and 2600 AD, according to the calculations by Meeus — was observed at many places and special expeditions were organised by Harvard College Observatory (6 stations) and the Smithsonian Astrophysical Observatory (2 stations).

The main results which concern us are the light curves obtained by DE VAUCOULEURS from H.C.O. observing at Péridier Observatory, Le Houga, France, and to a lesser extent at Boyen Observatory, Bloemfontein, South Africa, by ANDERSON. The whole observational material was discussed by MENZEL and DE VAUCOULEURS [5].

We shall give first of all a brief description of the observational method used at Le Houga, which was similar to that of Boyden Observatory. The photometry was performed with a stellar photometer placed in the focus of a 20 cm refractor. The photometer with 13-stage EMI multiplier under 1700 V was connected to the Brush recorder with a chart speed of about 21 mm/sec. The sky light, relatively strong at the moment of the occultation, was compensated by an auxiliary battery and the potentiometer. The net flux due to Regulus was about 0.3 V in comparison with the sky deflection of 3 V. The K-line (3933 Å) interference filter was used for the spectral selection.

As stated above (4.4) only the attenuation by refraction is to be taken into account. The light curve gives its value at each moment and using the Eq. (4.28) we may determine the value β of the density gradient in the homogeneous atmosphere

$$\beta = \frac{1}{H} = \frac{mg}{kT}. \qquad (4.32)$$

At both stations a systematic trend in the values of β has been found in the sense that β decreases with the altitude or the scale height H increases with it. The best solution for this variation may be written

$$H(h) = H(h_0)[1 + b(h - h_0)] \qquad (4.33)$$

with following numerical values

$$H(h_0) = 6.8 \pm 0.2 \text{ km}$$
$$b = +0.010 \pm 0.002 \text{ km}^{-1} = \frac{1}{H} \frac{\partial H}{\partial h}. \qquad (4.34)$$

The theoretical curve computed with the above assumptions is shown in the Fig. 4.6; the residuals are shown in the inset.

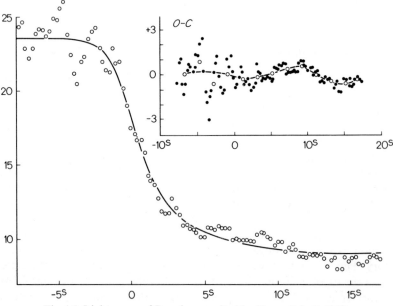

Fig. 4.6. Light curve of Regulus occulted by Venus July 7, 1959

With the acceleration of gravity $g(h_0) = 860 \text{ cm sec}^{-2}$, we get for the ratio

$$\frac{T}{\bar{m}} = 1.03 \, H \qquad (4.35)$$

or at the level where $H(h_0) = 6.8$ km we obtain

$$T = 7.0 \, \bar{m}. \qquad (4.36)$$

Further discussion depends on the assumption we make on the composition of the visible layer of the atmosphere. Adopting a different ratio $N_2:CO_2$, we get the following mean molecular mass and the temperature at the level h_0.

N_2/CO_2	1	$\frac{1}{2}$	0
m	36	40	44
$T(h_0)$	252	280	308

With the plausible value $m = 42.5$ the temperature would be $T(h_0) = 297° \pm 10°$ K.

Assuming the constancy of m with the altitude, i.e. the absence of any photodissociation, the value of b implies the temperature gradient

$$\frac{dT}{dh} = +3° \text{ km}^{-1}. \qquad (4.37)$$

On the contrary, if $T = $ const, the variation of the scale height implies the diminution of m with altitude connected to the photodissociation. Further speculations about the atmospheric model can be found in the obove quoted paper. For instance the air mass along the grazing ray at h_0 is only 1.3 m.

4.7. Occultation of σ Arietis by Jupiter, November 20, 1952

This occultation was photometered by BAUM and CODE [6] at the Cassegrain focus of the 60-inch telescope at Mt. Wilson. Owing to difficult observing conditions namely:

i) The difference of magnitudes between Jupiter -2.3^m and σ Arietis 5.5^m i.e. $\Delta m = 7.8^m$,

ii) blending of the star image with planet's limb,

iii) impossibility of using a sufficiently small focal diaphragme because of the scintillation,

the authors were compelled to resort to special measuring techniques. They took advantage of the difference between the reflected solar spectrum of Jupiter with the dark K-line and the B 5 spectrum of the occulted star where this line is nearly absent. The made use of a grating spectrograph for the necessary spectral selection. The entrance slit was replaced by a 1-mm round pinhole covered with the blue glass filter

BG-1. When bisected by the planet's limb, the pinhole admitted only 2% of Jupiter's light. This masking arrangement was combined with spectral selection of the spectrograph. The exit slit was centered on the K-line and open to about 10 Å. The relative gain to the photovisual region was 3 magnitudes. In this way the deflection due to the star before the occultation was nearly $\frac{1}{2}$ of the total deflection. The photometry was

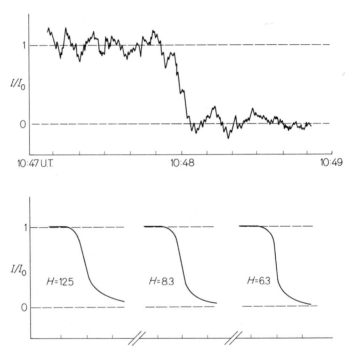

Fig. 4.7. Occultation of σ Arietis by Jupiter November 20, 1952. Above: The observed light curve. Below: The theoretical light curves

performed with the 1 P 21 multiplier and its current was recorded on a Brown Strip Chart Recording Meter.

For the interpretation of the light curve (Fig. 4.7) the authors start from the Eqs. (4.9)–(4.10) for the refraction

$$\omega = \omega_0 \exp -\frac{h_0}{H}, \quad \frac{d\omega}{dh_0} = -\frac{\omega}{H}$$

and from Eq. (4.23) for the attenuation by the refraction

$$\Phi = 1 + \frac{l}{H} \omega$$

which defines the light curve — the extinction being totally negligible. From a simple geometrical consideration (Fig. 4.8) we have also

$$\Phi = \frac{dy}{dh_0}. \qquad (4.38)$$

Now the time rate of Φ will be

$$\frac{d\Phi}{dt} = \frac{l}{H} \frac{d\omega}{dt}. \qquad (4.39)$$

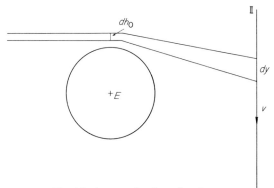

Fig. 4.8. Attenuation by refraction

Writing further for

$$\frac{d\omega}{dt} = \frac{d\omega}{dh_0} \frac{dh_0}{dt} = \frac{\omega}{H} \frac{dh_0}{dy} \frac{dy}{dt} = \frac{\omega}{H} \frac{1}{\Phi} v \qquad (4.40)$$

we have with the relative transverse velocity of the observer

$$v = \frac{dy}{dt}, \qquad (4.41)$$

where v is positive during ingress. Substituting for $\frac{l}{H} \omega$ from Eq. (4.23) and integrating, we obtain

$$\int_{\frac{1}{2}}^{\Phi} \frac{d\Phi}{\Phi^2 (\Phi - 1)} = \frac{v}{H} \int_0^t dt \qquad (4.42)$$

which yields finally

$$(\Phi - 2) + \log_e (\Phi - 1) = \frac{v}{H} t \qquad (4.43)$$

an equation of the family of light curves $\Phi = f(t)$ whose parameters are v and H.

Having plotted the curves $\Phi = f(t)$ for three values of $H = 6.3$; 8.3; 12.5 km, that of $H = 8.3$ seems to give the best fit. In the next step we can discuss according to the relation

$$H = \frac{kT}{mg} \qquad (4.32)$$

the mean value of the molecular weight m. Taking $g = 2600$ cm sec^{-2} and $T = 86°$ K in the Jovian stratosphere, we obtain $m = 3.3$. This value rules out the pure methane atmosphere ($m = 16$) and makes probable the mixture of H_2 ($m = 2$) and He ($m = 4$).

4.8. Near Occultation by the Planet

The next step in the exploration of the planetary atmosphere may be realized from the orbiting space probe enabling us to measure the light intensity of an occulted star. Planetary occultations from the Earth are extremely rare because of the small angular diameter of the planetary disk. From the planetary orbiter the planet can subtend an angle of several tens of degrees which considerably raises the probability of occultations.

In practice we may encounter occultations by Venus or Mars whose relatively thin atmosphere above the level of transparency permits us to use relatively simple formulae [13]. We have also for the density of the shadow

$$D = AM + \log\left[1 - (a+h)\frac{d\omega}{dh}\cos\psi\right] \qquad (4.44)$$

with the angle

$$\sin\psi = \frac{a+h_0}{a+h}. \qquad (4.45)$$

Using the formulae Eqs. (4.9), (4.29) we have calculated fictitious light curves for Venus and Mars [13] with the following parameters

	Venus	Mars
Distance of the probe $a+h$	18300 km	13200 km
$(a+h)\cos\psi$	17500 km	12430 km
H	6 km	17 km
Refraction at $h=0$	1'	2.76'
Extinction coefficient A	0.00575 km^{-1}	0.00575 km^{-1}

The corresponding light curves are represented in Fig. 4.9. In addition, we have assumed for Mars the existence of the thick dust layer up to the level $h_2 = 50$ km having the optical density in zenith $B = 0.03$. With

Eqs. (1.59) and (1.60) we have for the density along the horizontal trajectory

$$B(h_0) = 2B \sqrt{\frac{2a}{h_2}} \sqrt{1 - \frac{h_0}{h_2}}. \qquad (4.46)$$

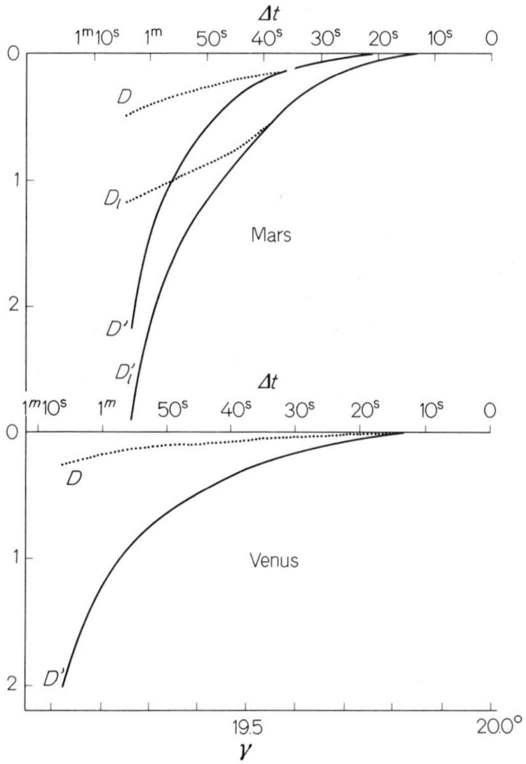

Fig. 4.9. Light curves for near occultations. D for the Sun, D' for the star in a pure atmosphere of Mars and Venus, D_l and D'_l the same with the dust layer for Mars

The light curves show clearly that during the occultation the rays can penetrate down to the planetary surface without being much attenuated. This circumstance, together with the relatively long time scale of the phenomenon, make it favourable for the photoelectrice photometry from the orbiter. The recent results obtained by the Mariner missions cannot modify this conclusion (6.8).

4.9. Eclipses of Phobos

Eclipses of Phobos, the inner satellite of Mars, were successfully photometered by RAKOS [7] using a special technique of photometrical scanning [8]. In fact, these measurements are extremely difficult when we realize that Phobos is fainter than Mars by a factor of 10^5 and at the moment of the eclipse only 4–5" in distance from the limb. The ordinary photometrical technique using a small diaphragm to diminish the diffused light is here ruled out by the smallness of the aperture necessary to attain it and the very high precision of guiding involved.

Therefore, the photometrical scanning was used in the focal plane of the 61-inch reflector of the U.S. Naval Observatory at Flagstaff. The slit of 50 microns in width and 500 microns in length was moved regularly from 1" to about 9" away from the Martian limb. In these absence of any object on the swept path the oscillographic record would give the

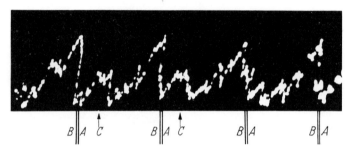

Fig. 4.10. Oscillogram of Mars and Phobos (C)

distribution of the scattered light in the vicinity of the planet (Fig. 4.10). In the presence of Phobos we obtain on the record a peak C signalizing both the position and the light intensity of the satellite (Fig. 4.10). Each point of the light curve was formed by the mean value of a large number of scans in order to reduce the statistical fluctuations of electrons and the effect of bad visibility. This last effect affects the observations much more indirectly by increasing the scattered light than directly by the broadening of the satellite image.

RAKOS succeeded in obtaining four eclipse curves in January and June 1965. The mean eclipse curve giving the loss of light in magnitudes plotted against the minimum height of the ray joining the satellite with the Sun's center ($=$ grazing height) is represented in Fig. 4.11.

For the theoretical interpretation, RAKOS followed in principle our theory. To take into account the limb darkening RAKOS uses the formula

$$b(R) = C(1 + \beta' \cos \gamma) \qquad (4.47)$$

which is equivalent to Eq. (1.16) if we put

$$C = 1 - \kappa, \quad C\beta' = \kappa, \quad \cos \gamma = \frac{1}{R_\odot} \sqrt{R_\odot^2 - R^2}.$$

His value of $C=\frac{1}{5}$ or $\kappa=2.8$ corresponds to the violet light and also to the curve of relative spectral response of the photometric system used. The solar disk was divided into 20 equal strips and the light intensity of each of them was computed by integration. For it he obtains the expression equivalent to our Eq. (2.11).

In the further development of the theory, RAKOS calculates in a very tedious manner the air mass $M(h_0)$, dividing the atmosphere into a number of concentric layers and adding the products of the paths in

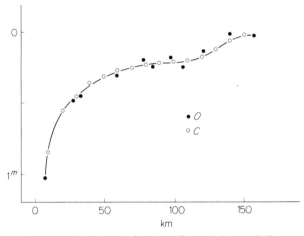

Fig. 4.11. Light curves of Phobos during the eclipse. O observed, C computed points

each layer and its mean air density. As he adopts the exponential structure of the atmosphere and the relatively small density at the surface, the classical formula [Eq. (4.9)] would bring him directly to the goal, but the use of an IBM 1620 computer compensated the slowness of his procedure and perhaps his overlook. The refraction is computed according to the Laplace theory from the air mass. In deriving the expression for the attenuation by the refraction, RAKOS closely follows our development used for lunar eclipses (1.2.10) with the final simplification due to the smallness of the refraction [Eq. (4.23)].

In the discussion of the mean light curve (Fig. 4.11) RAKOS pays considerable attention to the fall of light beginning at the height over 100 km. It is quite impossible to attribute this loss of light to the influence of any gaseous medium whoes density should be neglected at this level. He assumes therefore, the existence of solid particles at great heights in the Martian atmosphere.

By successive approximations with changing surface pressure and the distribution of solid particles with height, the computed loss of light was brought into good agreement with the observed light curve. Finally the adopted model of the atmosphere has the following parameters.

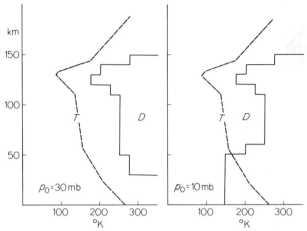

Fig. 4.12. Height distribution of aerosols (D) according to RAKOS, and for the temperature (T) according to CHAMBERLAIN

The distribution of solid particles in this model is given in Fig. 4.12. Its maximum almost coincides with the minimum of the temperature curve according to CHAMBERLAIN [9]. Some condensing process taking place in the coolest region of the Martian atmosphere may perhaps explain this peculiarity which presents some analogy with the terrestrial atmosphere.

Table 4.1

k (km)	10 mb model					30 mb model				
	F	kM	Ref.	G	Δm	F	kM	Ref.	G	Δm
10	0.214	0.197	0.079	0.339	0.83	0.151	0.501	0.162	0.196	1.00
20	0.010	0.135	0.053	0.361	0.56		0.362	0.116	0.213	0.69
30	0.000	0.059	0.023	0.363	0.45		0.188	0.058	0.241	0.49
40		0.023	0.009	0.340	0.37		0.090	0.027	0.252	0.37
50		0.009	0.003	0.307	0.32		0.043	0.013	0.260	0.32
60		0.003	0.001	0.271	0.27		0.020	0.007	0.253	0.28
70		0.001	0.000	0.249	0.25		0.010	0.003	0.244	0.26
80		0.000		0.232	0.23		0.005	0.001	0.233	0.24
90				0.222	0.22		0.002	0.000	0.221	0.22
100				0.213	0.21		0.000		0.213	0.21
110				0.200	0.20				0.200	0.20
120				0.175	0.17				0.175	0.17
130				0.126	0.13				0.126	0.13
140				0.073	0.07				0.073	0.07

Loss of light in stellar magnitudes: F by geometrical phase of the eclipse, kM by molecular scattering, Ref. by refraction, G by solid particles; Δm is the total loss of light.

However, this model is not definitive for surface pressure under 30 mb. RAKOS calculated another model with a surface pressure of 10 mb and a different distribution of solid particles (Fig. 4.12) which gives also good agreement with the observed light curve. In other words with the precision of measurements limited by the method used, we cannot distinguish between the two solutions. In the Table 4.1 are given the absorptions in both models separated into the atmospheric and solid particles components.

RAKOS concludes that the actual value of surface pressure can be determined only after the distribution of solid particles with height has been established. The field is open for further research with a more perfect technique on a greater number of eclipses.

4.10. Reduction of Eclipse Curves of Jovian Satellites

Let us first show Table 4.2 summarizing the geometrical eclipse conditions of old Jovian satellites.

Table 4.2

Satellite	R_S	π_S	n	$R_\odot$
I	0°.260	9°.748	0°.1413	0°.051
	2.66%	100%	1.45%	0.52%
II	0°.144	6°.092	0°.0704	0°.051
	2.37%	100%	1.16%	0.84%
III	0°.140	3°.817	0°.0349	0°.051
	3.66%	100%	0.915%	1.13%
IV	0°.077	2°.172	0°.0150	0°.051
	3.56%	100%	0.707%	2.35%

R_S and $R_\odot$ are angular radii of the satellite and the Sun observed from the Jupiter's center, π_S the parallax of the satellite and n its mean motion/min. All these numbers are also expressed in π_S as unity.

From this table we can see, for instance, that the disk of the satellites occupies in the shadow the angular interval $2R_S$ which cannot be neglected. As we can measure from the Earth only the global luminosity of the satellite, the immediate results of measurements may give only the average density of the shadow in the above interval. However, we are able to derive from the observed curve the reduced curve as if the satellite would have a negligible angular dimension.

The global luminosity of the satellite is given by the integral

$$e = \int T b \, dq \tag{4.48}$$

where T is the opacity of the shadow, and b the albedo of the satellite disk whose element is dq. The integral should be taken over the whole disk. Our task is to derive from the curve $e(t)$ obtained by observation as the function of the time t the opacity curve $T(t)$ which can be confronted directly with the theory [2].

The problem may be solved by successive approximations on the basis of assumed distribution of the albedo over the satellite disk. Putting at first $e = T_1$, we compute the above integral and we obtain the first approximation e_1 which is, of course, different from the actual value e. Modifying suitably the value from T_1 to T_2, we can repeat the evaluation of the integral to obtain the second approximation of e_2 and we continue until the difference $e_n - e$ becomes small enough with regard to t the accuracy of measurements.

In this procedure we may be considerably helped by knowing the mutual relations of the global curve e to the opacity curve T. Let us

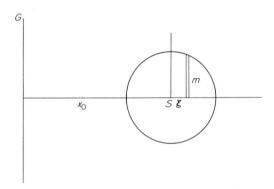

Fig. 4.13. Jovian satellite in the shadow

define the satellite position (Fig. 4.13) by its distance x_0 from the edge of the geometrical shadow G. In the small interval over the satellite disk the opacity of the shadow may be expressed by

$$T = K_0 + K_1 x + K_2 x^2 + K_3 x^3 + K_4 x^4 \tag{4.49}$$

as the isophotes of the shadow are practically rectilinear. The global luminosity of the satellite will then be

$$e = \int T f(\xi) \, d\xi \tag{4.50}$$

where $f(\xi)\,d\xi$ is the intensity of the elementar strip m (Fig. 4.12). Putting the value $x=\xi+x_0$ into Eq. (4.49), we obtain the above integral in the form

$$e = L_0 F_0 + L_1 F_1 + L_2 F_2 + L_3 F_3 \qquad (4.51)$$

where $L_0 = K_0 + K_1 x_0 + K_2 x_0^2 + K_3 x_0^3 + K_4 x_0^4$,
$L_1 = K_1 + 2K_2 x_0 + 3K_3 x_0^2 + 4K_4 x_0^3$,
$L_2 = K_2 + 3K_3 x_0 + 6K_4 x_0^2$,

$$F_0 = \int f(\xi)\,d\xi; \quad F_1 \int f(\xi)\,\xi\,d\xi = 0, \ldots, F_n = \int f(\xi)\,\xi^n\,d\xi.$$

The functions F_n depend only on the photometrical properties of the satellite disk and on his angular diameter $2R_S$. All these functions of odd orders are equal to zero because of the symmetry with regard to the center of the disk.

In the first approximation neglecting the first and higher powers of x, we arrive at the trivial relation

$$\frac{e}{F_0} = T \qquad (4.52)$$

already used above.

In the second approximation taking into account the second order terms we have

$$\frac{e}{F_0} = T_0 + L_2 \frac{F_2}{F_0} \qquad (4.53)$$

i.e. as long as L_2 is positive the global curve e will pass over the opacity curve T. In addition, whereas $L_2 = K_2 > 0$ we have also

$$\frac{d^2 T}{dx^2} = 2K_2 > 0 \qquad (4.54)$$

which is the criterion for the convexity of the opacity curve T with regard to the Ox axis. Everywhere the opacity curve is convex the global curve passes over it and vice versa (Fig. 4.13).

The slope of the global curve e is in the above approximation

$$\frac{1}{F_0}\frac{de}{dx_0} = K_1 + 2K_2 x_0 \qquad (4.55)$$

equal to that of the opacity curve T

$$\frac{dT}{dx} = K_1 + 2K_2 x$$

i.e. both curves are approximately parallel.

Going further to the third order terms, we get for the inflexion point of the global curve e

$$x = -\frac{1}{3}\frac{K_2}{K_3}, \quad \frac{e}{F_0} = T_0 \quad (4.56)$$

and equally for the opacity curve T, i.e. both curves cross mutually at their inflexion points.

Summing up, we may use these simple rules valid approximately for the mutual behaviour of the known global curve and the unknown opacity curve which are visualized in Fig. 4.14.

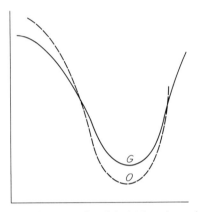

Fig. 4.14. Relation between the global (G) and opacity (O) curve

There remains only to derive the necessary expressions for the successive evaluations of the global luminosity integral [Eq. (4.50)] i.e. $f(\xi)$ and F_0. As far as the albedo b is concerned we have for instance the choice between the law of LOMMEL-SEELIGER or that of LAMBERT. At the opposition we have $b=1$ for the former law and $b=\cos \varepsilon$ for the latter with the incidence or diffusion angle ε. The needed expressions are, therefore,

$$f(\xi) = \sqrt{R^2 - \xi^2}, \quad F_0 = \frac{\pi R^2}{2} \quad \text{LOMMEL-SEELIGER}, \quad (4.57)$$

$$f(\xi) = \frac{\pi}{4R}(R^2 - \xi^2), \quad F_0 = \frac{\pi R^2}{3} \quad \text{LAMBERT}. \quad (4.58)$$

4.11. Eropkin's Eclipse Curves of Jovian Satellites

We shall now discuss Eropkin's measurements made by photographic photometry [10] which show some peculiar features on the eclipse curves. In fact on the decline of the curve beginning at an astonishingly great height we meet a kind of threshold (Fig. 4.15) with a small rise at the end and followed by a final drop to the zero intensity. These very curious characteristics expressed in heights are given in the following table.

Table 4.3

Satellite	I	II	III
Jovicentric latitude	+18°	+32°	+48°
Beginning of the decline	18%	11.5%	11%
	12800 km	8200 km	7900 km
Secondary minimum	10%	6%	5%
	7100 km	4300 km	3600 km

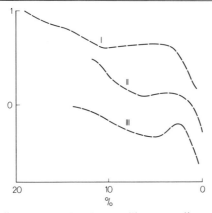

Fig. 4.15. Eclipse curves of Jovian satellites according to EROPKIN

One might be tempted to ascribe the early beginning of the decline to the attenuation by refraction in a very rarefied atmosphere. Assuming an attenuation of 0.99 at the level of 13000 km within the exponential atmosphere having the plausible parameters at this height

$$\beta = 3 \times 10^3, \quad \rho = 10^{-8}$$

we obtain for the surface density a totally absurd value of $\log \rho = 230$. In other words if the density at the surface does not exceed some reasonable, even very great value $\rho_0 < 1000$, the attenuation at 13000 km will be negligible. These reasons lead us [2] to put forward the hypothesis of a dust layer surrounding the planet outside its atmosphere and forming there some kind of a rudimentary ring like the Saturn crepe ring.

Recently HARRIS [11] published some photoelectric measurements which do not agree with Eropkin's curves. It would perhaps be necessary to investigate the reasons for this disagreement on the basis of more observational material before we conclude about the possible light absorption in the proximity of Jupiter.

4.12. Frost Phenomena on Jovian Satellites during Their Eclipses

During the eclipse, whose length may be 2−3 hours, the satellites receive only a very weak long-wave radiation from Jupiter's night hemisphere and the surface temperature must be of about 100 °K. BINDER and CRUIKSHANK [12] have, therefore, examined the possibility of some condensation-frost or haze, visible just after the exit from the shadow

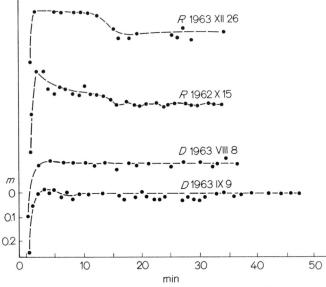

Fig. 4.16. Light curves of Jovian satellite I ($=Io$) during the disappearance (D) and the reappearance (R)

before the solar radiation had restored the normal temperature condition. The condensations causing an albedo increase may involve also the light increase at the reappearance which would decay to the normal light some time later and this phenomenon may be detectable by photometry.

Several series of photoelectric measurements were performed from 1962 to 1965 at the Steward Observatory and at the Kitt Peak National Observatory in order to obtain good eclipse curves. The first satellite J-I displayed at the reappearance a light increase with the maximum of 0.09^m and of 15 min duration. The test curves at the disappearance have shown no effect of this kind (Fig. 4.16). The second satellite J-II showed once an increase of 0.03^m lasting 10 min. The third satellite J-III failed to show any anomaly.

From a purely photometric point of view we may discuss for J-II two or three factors involved in the problem, i.e. the normal satellite albedo, the albedo of the condensation product and the fraction of surface covered by it just at the reappearance. If the snow albedo 0.8 is considered, the 25% surface cover gives an increase of 0.12^m. Besides the photometrical side, we must consider also physical conditions. The thermal escape of gases, and their pressure at the given temperature must be considered in this context. Two gases, NH_4 or N_2, have some chance of being available in significant amounts. For J-II satellite the conclusions are uncertain, and for J-III high quality light curves showing no anomaly tentatively exclude any appreciable atmosphere of these gases.

4.13. Terrestrial Occultation Observed from the Moon

Knowing the structure of the auxiliary shadow during the lunar eclipses (1.2.11), we are able to discuss the possible occultations of stars by the Earth as observed from the Moon. As the width of the zone swept

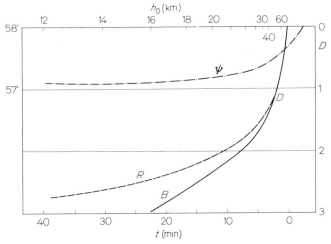

Fig. 4.17. Terrestrial occultation of a star as observed from the Moon: ψ apparent distance, and the density D in red (R) and blue (B). Above the height and below the time scale

on the sky by the Earth during its apparent revolution (27 days) round the Moon is about $3\frac{1}{2}$ times larger than that of the Moon seen from the Earth the probability of occultation on the Moon is greater in the same ratio. In addition, the presence of the Sun above the lunar horizon could not inhibit these observations. The lunar observer or some automatic device may, therefore, measure the course of the terrestrial occultations.

Assuming the central occultation with Earth's hourly motion of 0.5°, we have drawn the light curve of an occultation in red and blue light using the data of Table 1.9. We see that the duration of occultation up to the total disappearence at $h_0 = 2$ km (cloud level) is about two hours, but (Fig. 4.17) the best observable phase from the beginning to the drop down to 1% of the initial intensity will last about 8 min. This long duration would largely facilitate eventual measurements on the Moon. As the apparent motion of the occulted star is concerned the curve of ψ (Fig. 4.17) shows the same feature as the phenomena of far occultation i.e. the long attachement of the star image to the Earth's limb during a major part of the phenomenon.

The geographic differences of the atmospheric structure will appear, of course, in the form of the light curve as well as the differences of the Rayleigh's atmosphere as those of other atmospheric constituents (O_3 or dust).

Bibliography

1. SCHMIDT, A.: Die Strahlenbrechung auf der Sonne. Stuttgart 1891.
2. LINK, F.: Bull. astron. **9**, 227 (1936).
3. PANNEKOEK, A.: A.N. **164**, No. 3913 (1903).
4. FABRY, CH.: J. Observ. **12**, No. 1 (1929).
5. MENZEL, D.H., and G. VAUCOULEURS: Final Report, AFCRL-227 (1961).
6. BAUM, V.A., and A.D. CODE: Astron. J. **58**, 108 (1953).
7. RAKOS, K.D.: Lowell Obs. Bull. No. **131** (1966).
8. RAKOS, K.D.: Appl. Optics **4**, 1453 (1965).
9. CHAMBERLAIN, J.: Aph. J. **136**, 582 (1962).
10. EROPKIN, D.J.: Z. Aph. **3**, 163 (1931).
11. HARRIS, D.J.: Planets and Satellites in the Solar System (eds. KUIPER-MIDDLEHURST). Chicago 1961.
12. BINDER, A.B., and D.P. CRUIKSHANK: Icarus **3**, 299 (1964); **5**, 7 (1966).
13. LINK, F.: Space Research **3**, 1022 (1963).

5. Transits of Planets over the Sun

5.1. Introductory Remarks

Transits of planets over the Sun became a part of astronomy mainly as the transits of Venus, which have been used, according to the proposals outlined by HALLEY, for the determination of the solar parallax. Much later INNES used the transits of Mercury for some special investigations about the Earth's rotation.

The four transits of Venus in the years 1761, 1769, 1874 and 1882 provided extensive observational material collected by numerous astronomical expeditions sent to all parts of the Earth [1]. Fortunately enough, in the huge mass of observations mainly of astrometrical character some unusual features, such as the aureola around the Venus disk, were noticed and their interpretation gave, early in the 19th century, the first indications of the planetary atmosphere.

From the formal viewpoint, the transits belong to the same chapter as the observations of lunar eclipses on the Moon (1.2). However, there is a very important difference in the relative size of the dark occulting planet and the Sun. During lunar eclipses the Earth's dark disk is about $4\times$ larger than the Sun while during the transits the Sun is $32\times$ larger than Venus. Therefore, the conditions for observations and the course of the phenomena are quite different in both cases.

5.2. Refraction in the Planetary Atmosphere

For the terrestrial observer N (Fig. 5.1) Venus is projected on the solar plane I. A solar point M is seen from the Earth at the angle ψ and its position on the solar plane is fixed by the angle r. The relation

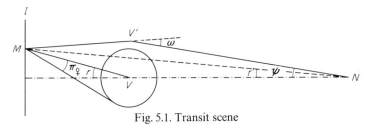

Fig. 5.1. Transit scene

between the object angle r and the image angle ψ is like that for lunar eclipses [Eq. (1.11)]

$$r = \pi_{\venus}\left(1 + \frac{h'_0}{a}\right) + \psi - \omega \tag{5.1}$$

where $\pi_{\venus}$ the angular radius of Venus as seen from the Sun or the planetary "parallax"
$r_{\venus}$ the angular radius of Venus as seen from the Earth
h'_0 the altitude of the asymptote of the ray
a the radius of Venus
ω the deviation of the rays due to refraction.

In mean transit conditions it will be (Fig. 5.1)

$$r' = r\frac{r_{\venus}}{\pi_{\venus} + r_{\venus}} = 0.725\, r. \tag{5.2}$$

Further we have as for lunar eclipses [Eq. (1.10)]

$$\psi = r_{\venus}\left(1 + \frac{h'_0}{a}\right). \tag{5.3}$$

The two formulae [Eqs. (5.2)–(5.3)] enable us to construct the solar image formed by the refraction in the planetary atmosphere.

5.3. Refraction Image of the Sun

We shall first of all examine the *limiting curve* in the solar plane, i.e. the curve which separates the visible and the invisible parts of the solar disk as seen from the Earth. This curve is defined by the ray tangent to the visible planetary surface (cloud level) where $h'_0 = 0$ and ω_0 is the maximal value of the refraction at the same level. Putting these values into Eq. (5.1) we get

$$r_0 = \pi_{\venus} + r_{\venus} - \omega_0 \tag{5.4}$$

or for the terrestrial observer

$$r'_0 = r_{\venus} - 0.725\, \omega_0 \tag{5.5}$$

the equation of the limiting curve.

Let us now consider the form of the limiting curve along the planetary limb where the refraction may vary to some extent. As long as $\omega_0 < 43'' = \pi_{\venus} + r_{\venus}$, the angles r_0 and r'_0 are *positive* and this part of the limiting curve should be denoted as the *positive part*. In the case when $43'' < \omega_0$, these angles become *negative* and this part of the limiting curve is called the *negative part*.

As the value of limiting refraction $\omega_0 = 43''$ separating the two regimes of the positive and negative parts of the limiting curve is small, we can meet both along the limb of Venus with a positive part in equatorial regions and a negative part near the poles with a transition in temperate latitudes. This general case is illustrated by a fictions example given in Fig. 5.2. The limiting curves, positive (fully drawn) and negative (dashed), have here a form of two crossed eights with their nodal points at temperate latitudes corresponding to $\omega_0 = 43''$.

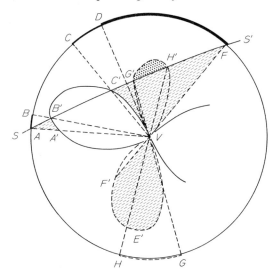

Fig. 5.2. Formation of solar image

The image of the solar plane visible from the Earth is constructed according the following rules:

1. The points of the solar plane lying inside the positive lobes are invisible from the Earth and vice versa.

2. The points lying inside the negative lobes are visible from the Earth and vice versa.

For a given phase of the transit the solar border occupies the position SS' (Fig. 5.2). Then we have the solar image composed of following parts of the aureole:

SB being the image of the solar area $AA'B$ lying outside the positive lobe I_p (rule 1).

CS' being the image of the solar area $C'FV$ lying between the positive lobes I_p and II_p.

DS' being the image of the solar area $VE'F'V$ lying inside the negative lobe IV_n.

The dark protrusion HG being the image of the part $G'H'$ of the negative lobe II_n lying outside the solar disk.

It remains to determine the thickness or the radial extension of the aureola. We have for this the difference of angles ψ corresponding to the angles r_0 on the limiting curve and r_1 on the solar border. According to Eq. (5.3) we have

$$\Delta\psi = \psi_1 - \psi_0 = r_\varsubsun(h'_1 - h'_0). \tag{5.6}$$

As we may assume the exponential structure of the outer parts of the planetary atmosphere, we can write [Eq. (4.9)]

$$\omega_1 = \omega \exp - \beta h_1$$
$$\omega_0 = \omega \exp - \beta h_0 \tag{5.7}$$

and their ratio

$$\frac{\omega_0}{\omega_1} = \exp \beta(h_1 - h_0) \tag{5.8}$$

gives the difference of the altitudes

$$h_1 - h_0 = \frac{2.30}{\beta} \log \frac{\omega_0}{\omega_1}. \tag{5.9}$$

It follows further from Eq. (5.4)–(5.5)

$$\omega_0 = (\pi_\varsubsun + r_\varsubsun)\left(1 + \frac{h_0}{a} - \frac{r'_0}{r_\varsubsun}\right)$$
$$\omega_1 = (\pi_\varsubsun + r_\varsubsun)\left(1 + \frac{h_1}{a} - \frac{r'_1}{r_\varsubsun}\right) \tag{5.10}$$

or for the difference

$$h_1 - h_0 = \frac{2.30...}{\beta} \log \frac{1 + \dfrac{h_0}{a} - \dfrac{r'_0}{r_\varsubsun}}{1 + \dfrac{h_1}{a} - \dfrac{r'_1}{r_\varsubsun}} \tag{5.11}$$

and finally the desired difference

$$\Delta\psi = \frac{2.30...}{a\beta} \log \frac{r_\varsubsun - r'_0 + r_\varsubsun \dfrac{h_0}{a}}{r_\varsubsun - r'_1 + r_\varsubsun \dfrac{h_1}{a}} \approx 2.30... \frac{r_\varsubsun}{a\beta} \log \frac{r_\varsubsun - r'_0}{r_\varsubsun - r'_1} \tag{5.12}$$

where the differences

$$r_\varsubsun - r'_0; \quad r_\varsubsun - r'_1$$

may be read directly in Fig. 5.2.

The product $a\beta$ is of the order of 10^3 and, therefore, the thickness of the aureole will be less than $1''$, i.e. on the limit of the resolving power of the observing instruments. Its brightness will be in consequence

$$l = \text{const } b \, \Delta\psi \qquad (5.13)$$

i.e. proportional to the thickness of the aureola and to the mean surface brightness of the projected part of solar disk.

5.4. Discussion of Past Transits

We shall now discuss the observations made during the past four transits. As it would be quite impossible to give here the texts of the observations in full, we have extracted from them the most important information and arranged it under the following items:

a) The existence of the aureole.

b) The prevalence of its intensity towards N or S.

c) The closure or disruption of the aureole.

d) The appearance or disappearance of the aureole.

e) The formation of the cap.

f) The visibility of light spot.

g) The visibility of the dark planetary disk outside the solar disk and surrounded by a faint glow.

For each of the above phenomena, if observed, we give the corresponding phase $1 > \varphi > 0$ equal to the fraction of the planetary diameter outside the solar disk. A total of 94 observations of the aureole has been collected from reports on the transits in 1961, 1769, 1874 and 1882. Their characteristics described with the aid of the above scheme are given in the table.

Table 5.1

	a	b	c	d	e	f	g
Entry 1761	0.17 (2)	—	—	—	—	—	—
Exit 1761	0.15 (3)	—	0.30 (2)	0.67 S (1)	0.75 S (1)	—	—
Entry 1769	0.10 (1)	—	—	0.55 (7)	—	0.67 N (1)	—
Entry 1874	0.29 (10)	0.30 (3)	0.38 (7)	0.58 (2)	—	—	0.98 (4)
Exit 1874	0.40 (1)	0.34 (6)	0.50 (6)	0.50 (1)	0.81 (6)	0.70 (4)	—
Entry 1882	0.23 (12)	0.37 (4)	0.46 (11)	—	0.62 (7)	0.55 (2)	0.56 (1)
Exit 1882	0.10 (1)	0.22 (4)	0.45 (5)	0.55 (1)	0.65 (2)	—	0.70 (1)

The table gives for different transits and different forms of the aureola $a-g$ the mean phase and the number of observations.

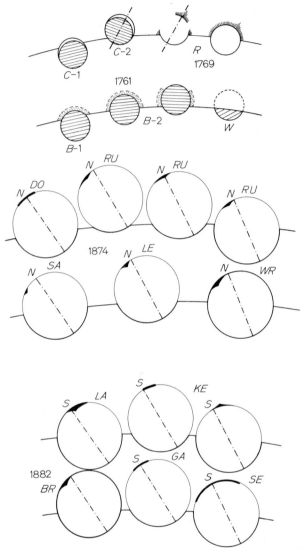

Fig. 5.3. Different observations of the aureole during the transits. 1761 transit: $C-1, C-2$ CHAPPE D'AUTEROCHE; $B-1, B-2$, BERGMANN; W WARGENTIN. 1769 transit: R RITTENHOUSE. 1874 transit: Do DÖLLEN; Ru RUSSEL; Sa SAVAGE; Le LENEHAM; Wr WRIGHT. 1882 transit: La LANGLEY; Ke KEELER; Br BRASHEAR; Ga GARNIER; Se SEELIGER

Let us add some more detailed quotations of the most interesting observations.

1761.

CHAPPE D'AUTEROCHE has noted an aureole which he calls "a little atmosphere" (Fig. 5.3) and was able to follow its last remnants detached from the solar limb as far as the phase $\varphi = 0.75$. Several other observers have also observed aureoles. Lomonosov's observations will be discussed later.

1769.

RITTENHOUSE has noted at the ingress a light pyramid (Fig. 5.3) lowering with the progress of the phenomenon and widening around the planetary limb. At the bisection ($\varphi = 0.5$) the aurole was already complete with the strengthening in the direction of the pyramid.

1874.

Russell's series of images during the egress gives the complete aureole as far as the phase $\varphi = 0.45$ when the disruption takes place leaving the polar cap. The cap after the progressive contraction disappears at $\varphi = 0.97$. Similar feautures have been reported by SAVAGE, LENEHAM and WRIGHT (Fig. 5.3).

1882.

KEELER noted first at the ingress at $\varphi = 0.75$ the light spot which extended to the cap. At the bisection ($\varphi = 0.5$) the aureole was complete with the remaining light spot. The position of this spot was confirmed by LANGLEY, SEELIGER and BRASHEAR (Fig. 5.3).

5.5. Simplified Discussion of the Transit Phenomena

For the discussion of the observations it is useful to reduce the general discussion given above to a simplified form of two limiting cases.

A. The limiting curve is a positive circle for $\omega_0 < 43''$. The case is represented in the upper part of Fig. 5.4 for the egress. As long as the limiting circle is inside the solar disk we observe a complete aureola discernible, however, only along the external part of the planetary limb. The disruption of the aureola takes place at the moment of the internal contact of the limiting circle with the solar border. Afterwards the aureola splits into two corners adjacent to the solar border and shrinks progressively until the disappearance at the moment when the planet is bisected by the solar border.

B. The limiting curve is a negative circle for $\omega_0 > 43''$ (Fig. 5.4 lower part). At the beginning there is no difference from the previous case until the bisection moment when the aureola detaches itself from the solar border and continues to be visible as a cap. The cap shrinks gradually and disappears at the moment of the external contact of the limiting circle with the solar border, no matter what is the actual position of the planetary limb (Fig. 5.4), i.e. also when the planet has already left the solar disk ($|r'_0| > r_♀$).

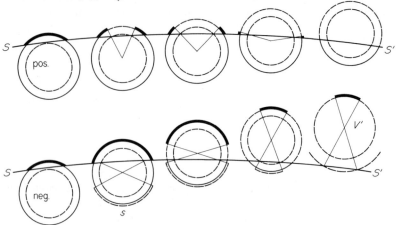

Fig. 5.4. Simplified scheme of the transit for positive and negative limiting curves, V' the case when $|r'_0| > r_♀$

This discussion leads us to a simple method for the analysis of the observations. At the moment of the disruption in the positive case or at the complete disappearance in the negative case the refraction is given by

$$\omega_0 = (\pi_♀ + r_♀)\left(1 - \frac{\Delta}{r_♀}\right) \qquad (5.14)$$

where Δ is the distance of the planet center from the solar border positively counted towards the interior.

In the negative case, the existence of the aureola after the bisection, i.e. the existence of the negative case itself, leads to the lower limit for the refraction

$$\omega_0 > \pi_♀ + r_♀. \qquad (5.15)$$

From the observational material discussed, the existence of the caps during each transit is evident. They mark the spots on the planetary limb where the refraction is greatest and in consequence the temperature lowest, i.e. the polar regions. At its disappearance the position of the planet Δ gives with the aid of Eq. (5.14) the amount of the refraction at the poles. On the sketches drawn by different observers we may in addition determine the position angle P of the poles. Both quantities are given in the following table.

Simplified Discussion of the Transit Phenomena

Table 5.2

Transit	Cap	LINK [1]			KUIPER [12]		Season
		P obser- vations (°)	Number of obser- vations	Refrac- tion	P (°)	p (°)	
1761	South	200	1	1.1?	177	120	Winter
1769	North	350	3	1.1	357	60	Summer
1874	North	20	3	1.7	3	120	Winter
1882	South	180	3	1.1	183	60	Summer

Furthermore we may compare these indications with the position of the rotation axis determined by KUIPER [2] from the ultraviolet photographs of the surface where a system of parallel bands seems to indicate the direction of the equator. When α_0 and δ_0 are the equatorial coordinates of the Venus north pole, the position angle of rotation axis P as seen from the Earth and the polar distance p of the Sun on Venus are given by

$$\sin P = \frac{\sin(\alpha_\venus - \alpha_0)}{\sin p} \cos \delta_0$$

$$\cos p = \sin \delta_0 \sin \delta_\venus + \cos \delta_0 \cos \delta_\venus \cos(\alpha_\venus - \alpha_0). \tag{5.16}$$

The values of P and p with the corresponding seasons on Venus are also given in the above Table 5.2. The agreement of both sets of data is as good as possible considering the quality of the observations.

Summing up, we may assert that there exist on Venus polar and equatorial regions with low or high temperatures and that the temperature variations agree with the orientation of the rotation axis towards the Sun. These facts are incompatible with the hypothesis of the equality of the revolution and the rotation of the planet. If this were so, Venus would turn the same face to the Sun and no temperature differences along the limb could exist nor would they be observed during the transits. The rotation in respect to the Sun must be, therefore, different from the revolution around it. This conclusion was indeed confirmed by the last visual or radar observations even though these observations disagree somewhat on the proper length of the rotation.

The values of the refraction concluded from the aureola

Equator	Pole-Summer	Pole-Winter
0.7′	1.1′	1.7′

may lead with the aid of Eqs. (4.2) and (4.9)

$$\omega = A \frac{p}{\sqrt{T^3}}, \quad A = \begin{cases} 393' & \text{for } N_2 \\ 741' & \text{for } CO_2 \end{cases} \quad (5.17)$$

to some information about the temperature at the cloud level, however, only if we admit a certain composition of the atmosphere. The graphs of Eq. (5.17) are given in the Fig. 5.5 for two gases which possibly exist on Venus, namely CO_2 and N_2, the first being disclosed by spectrography.

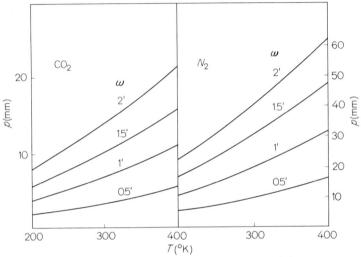

Fig. 5.5. Refraction on Venus as function of the pressure and the temperature for two compositions of the atmosphere

Unfortunately the problem is not here so simple, for even if the composition is adopted, we still have the choice of the pressure for a given temperature and vice versa.

As far as the refraction variations along the limb are concerned, they require the relatively great variations of temperature lying between 200 and 400° K. We must also keep in mind that the origin of the refraction variations may be somewhat more complex. If there really exist temperature variations at the level of the transparency (=clouds) this level itself may vary with the temperature as we know it on Earth. In other words the small fall in temperature from the equator to the poles makes the level of the transparency drop and in consequence increases the refraction. We can show that a decrease of a few kilometers doubles the air density and in consequence the refraction. The conclusion is perhaps that by discussion the above graphs (Fig. 5.5), we must move there in a vertical rather than in a horizontal direction between the equator and the poles.

5.6. Lomonosov's Phenomenon

Lomonosov's observations of the 1761 transit merit some special remarks not so much for their scientific value, but for the unfounded appreciation by SHARONOW [3]. LOMONOSOV [4] observed this transit with a small telescope ($4\frac{1}{2}$ feet long) of bad optical quality giving colored images outside the optical axis. Before the egress, when the limb of Venus was at a distance by 1/10 of its diameter from the solar limb, LOMONOSOV detected at the solar limb a kind of swelling or blister becoming more distinct as the planet was approaching the egress. A short time afterwards the blister disappeared and the planet was visible without any special feature at the solar limb (Fig. 5.6). LOMONOSOV explained this phenomenon by the refraction in the planetary atmosphere which is, according to him, equal if not more important than the terrestrial atmosphere.

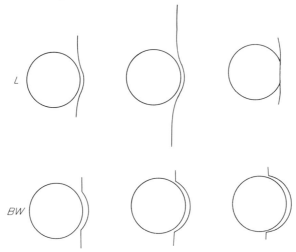

Fig. 5.6. Lomonosov's phenomenon observed by himself in 1761 and by BIGG-WITHER in 1874

113 years later the British astronomer BIGG-WITHER [10] observed a very similar phenomena at the egress. When Venus was approaching the egress the planet seemed to push before it a kind of light ring. This feature was observed at the moment of the computed internal contact. Soon after, when the disk was outside the Sun, the ring in the form of a crescent was visible (Fig. 5.6).

Both Lomonosov's and Bigg-Wither's phenomena show a remarkable resemblance to each other and have probably the same origin in the irradiation which takes place on the limit of two areas of very different brightness observed under bad conditions. In addition, the explanation given by LOMONOSOV is not valid as it presupposes the refracting atmosphere at the height of 1/10 of diameter (= 1200 km), which is impossible, thus making it clear that LOMONOSOV never saw an atmospheric phenomenon which is different from the appearance observed by him and BIGG-WITHER.

We feel 1° that LOMONOSOV proposed though somewhat incautiously an explanation by atmospheric refraction for the phenomenon observed by him which was, however, not of atmospheric origin, and 2° that his contemporaries having observed the true atmospheric phenomenon proposed independently but with more caution the same explanation.

Neither LOMONOSOV nor any other astronomer has established (1761) any theory or brought to light any evidence to support their findings. The schematic outline of refraction presented by LOMONOSOV cannot, therefore, be considered as a form of theory and if it should be according to SHARONOW, then this theory will have led to false conclusions about the importance of the Venusian atmosphere as quoted above. Hence we may admit under the denomination of Lomonosov's phenomenon only for the irradiation blister observed by him and later by BIGG-WITHER and not for the true aureole observed in 1761 by nearly a dozen other astronomers. Thus SHARONOW, having collected and published a very copious documentation [3] about Lomonosov's contribution, has not convinced the present author of its scientific value, neither has it convinced STRUVE who, some years ago, carried out an independent investigation [5]; he, too, came to similar conclusions.

5.7. Extension of the Cusps of Venus

The phenomenon known under this name was observed for the first time at the end of 18th century and was ascribed by several authors to the refraction in the Venusian atmosphere. The real origin is, however, the scattering of solar rays in the hazy atmospheric layers as it will be demonstrated further. The phenomenon is visible near the transits and, therefore, we shall pay them some attention, the more so as its theory has some points in common with transit theory.

As Venus is approaching the inferior conjunction, the very narrow crescent of the planet has the central angle $180° + 2p$ instead of the theoretical value of $180°$ (Fig. 5.7). The reason for the extension p can be understood geometrically as the consequence that the shadow terminator s as well the visibility terminator t from the Earth are shifted beyond their theoretical positions (Fig. 5.7). Thus the cusp of the crescent is not at the point V but in the intersection of s with t at the point C. From the spherical triangle $E'S''C$ we get

$$\sin \sigma = -\sin \tau \cos \varphi + \sin \varphi \cos \tau \sin p \qquad (5.18)$$

Extension of the Cusps of Venus

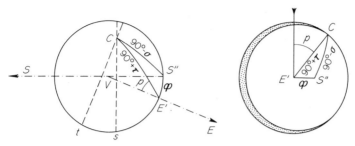

Fig. 5.7. Extension of the cusps of Venus

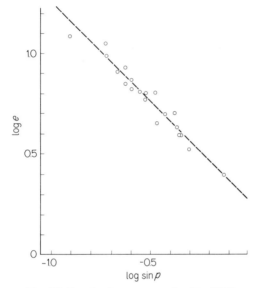

Fig. 5.8. Graph of cusps extension Eq. (5.21)

where the phase angle φ is computed from the elongation e of the planet from the Sun by means of

$$\sin \varphi = \frac{\pi_\varphi + r_\varphi}{r_\varphi} \sin e. \tag{5.19}$$

The angle involved being small we can simplify to the form

$$\frac{\pi_\varphi + r_\varphi}{r_\varphi} e \sin p = \tau + \sigma = q \tag{5.20}$$

or in the logarithmic form suitable for graphic representation

$$\log q = \log e + \log \sin p + 0.14. \tag{5.21}$$

In Fig. 5.8 the observation collected in the Table 5.3 are plotted in order to determine the value of [6]

$$q = 2.63° \pm 0.05°. \tag{5.22}$$

According to RABE, the value of q depends a little on the aperture of the observing instruments and on the atmospheric conditions. RABE finds these limits $1.9° < q < 3.7°$ in good agreement with the above given average value of $2.6°$.

Table 5.3

	Date		e	p	$\log e$	$\log \sin p$	$\log q - 0.14$
MADLER	1849 V	10	5° 08'	22°½	0.71	−0.42	0.29
		11	3° 57'	27°½	0.60	−0.34	0.26
		11	3° 36'	27°½	0.53	−0.30	0.23
		12	3° 26'	30°	0.53	−0.30	0.23
		15	6° 25'	17°½	0.81	−0.52	0.29
		16	7° 37'	15°	0.88	−0.59	0.29
LYMAN	1866 XII	7	6° 25'	20°	0.81	−0.47	0.34
		14	5° 06'	25°	0.71	−0.37	0.34
		15	6° 43'	15°	0.83	−0.59	0.24
		18	11° 23'	11°	1.06	−0.72	0.34
	1874 XII	8	2° 32'	49° 41'	0.40	−0.12	0.28
		11	4° 02'	26° 38'	0.60	−0.35	0.25
		11	4° 20'	25° 43'	0.64	−0.36	0.28
		12	5° 58'	17° 41'	0.78	−0.52	0.26
RABE	1913 IV	21	8° 18'	12.4°	0.92	−0.66	0.26
		22	7° 14'	13.9°	0.86	−0.62	0.24
		26	6° 36'	14.6°	0.82	−0.55	0.27
		28	8° 40'	14.0°	0.94	−0.62	0.32
		29	9° 57'	11.1°	1.00	−0.71	0.29
	V	1	12° 36'	7,2°	1.10	−0.90	0.20

5.8. Explanations of the Cusps Extension

The above relations assume nothing about the true origin of both shifts of the terminators. At first the refraction was taken into account for this purpose. From Fig. 5.9 we get

$$\sigma = \frac{\omega}{2} - \pi_\venus + r'_\odot$$

$$\tau = \frac{\omega}{2} - r_\venus \tag{5.23}$$

where

$$r'_\odot = R_\odot \frac{\pi_\venus + r_\venus}{r_\venus} \tag{5.24}$$

is the solar apparent radius as seen from Venus.

The refraction ω can be computed from the sum q

$$\omega = q + \pi_{\venus} + r_{\venus} - r'_{\odot} \tag{5.25}$$

and we get with $q = 2.65°$ the minimum value of $\omega = 2.25°$ since the observable penumbra on the shadow terminator will be surely narrower than $r'_{\odot}$.

This value is then considerably greater than that obtained from the transits ($\omega = 1'$). We cannot explain this disagreement by the difference of levels at which the grazing rays penetrate in both phenomena. Due to the great brightness of solar disk, this level should be lower and the corresponding refraction greater by the transit than by the inferior conjunction, where the intensity of images is considerably less.

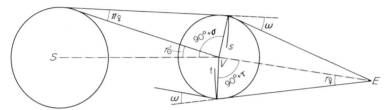

Fig. 5.9. Explanation of the cusps extension

Besides this discrepancy, two very conclusive arguments have been advanced against the role of the refraction in the cusps extension. The first was put forward by RUSSELL [7] and it is based on the absence of the refraction image at the moment of the closing ($p = 90°$) of the cusps. Introducing $p = 90°$ into Eq. (5.20), we obtain the condition

$$e = \frac{r_{\venus}}{\pi_{\venus} + r_{\venus}} (\omega - \pi_{\venus} - r_{\venus} + r'_{\odot}) = R_{\odot} - \Delta \tag{5.26}$$

or after some transformations

$$\omega = (\pi_{\venus} + r_{\venus})\left(1 - \frac{\Delta}{r_{\venus}}\right) = (\pi_{\venus} + r_{\venus})\left(1 - \frac{R_{\odot} - e}{r_{\venus}}\right) \tag{5.27}$$

which is identical with Eq. (5.14) determining the moment of the first appearance (at the ingress) of the refraction image. Therefore, with closing cusps the refraction image should appear at the point of the closure. The observations showed, however, that the closing of cusps takes place at the elongation $e = 1.9°$ whereas the first refraction image appears much later at $e = 0.25°$. The gap between the two events shows necessarily their different origin.

The second argument [6] is the color of the extended cusps. The refraction and the air mass traversed by the ray are roughly proportional (1.2.22), therefore, with the refraction of $2°$ the air mass of several hundred kilometres should be expected. Such a great air mass has as a consequence not only a very large attenuation of the light but also its deep red color, which has never yet been observed.

SCHOENBERG remained an adherent of the ancient refraction theory and calculated the atmospheric structure based on this assumption [8]. To dispose of Russell's argument, he put forward a somewhat curious explanation: the refraction image cannot be visible as it is blurred by the effect of the chromatic dispersion [9]. A white ray of light during its passage through the Venusian atmosphere is dispersed into a spectral

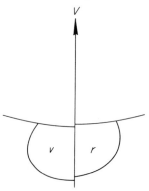

Fig. 5.10. Violet and red solar images according to the Schoenberg's hypothesis

fan whose width at the Earth's distance is of several millions kilometres. Consequently the intensity of the ray is reduced at such a rate that it becomes invisible from the Earth.

There is, however, an error in the principle of his argument. A simple ray cannot transport any energy, we must always consider the existence of even a very narrow pencil of rays. The attenuation of the light transported by this narrow pencil is given by Eq. (1.35). The dispersion of light, i.e. the dependence of the refraction on the wave length is here of minor importance [9].

We can visualize this conclusion by constructing solar refraction images in violet and red light adopting the structure of Venusian atmosphere according to Schoenberg's overestimated parameters. With it we calculated the following Table 5.4 which is self-explanatory in the context with the above given formulae. From this table we may trace the red and violet images at the presumed closing of cusps (Fig. 5.10). We see that there is only a small difference of sizes at both ends of the visible spectrum. In other words the white image of the Sun at the critical moment as well as at any other moment during the transit will be only slightly smeared by the dispersion. In any case, if it really exists, it should be observable from the Earth without any appreciable attenuation.

Table 5.4

h_0 (km)	Red				Violet			
	ω (')	h'_0 (km)	$\Delta h'_0$ (km)	$\Delta \psi_0$ 0.001"	ω (')	h'_0 (km)	$\Delta h'_0$ (km)	$\Delta \psi_0$ 0.001"
0.0	131.8	3.35	0.00	0.00	134.4	3.42	0.00	0.00
0.5	123.7	3.64	0.29	1.47	126.2	3.71	0.29	1.47
1.0	116.1	3.95	0.60	3.05	118.4	4.00	0.58	2.94
1.5	109.0	4.27	0.92	4.67	111.2	4.33	0.91	4.62
1.7	106.3	4.40	1.05	5.34	108.1	4.45	1.03	5.23
1.9	103.6	4.53	1.18	6.00	105.7	4.55	1.13	5.74
2.0	102.3	4.60	1.25	6.34	104.3	4.65	1.23	6.25
2.1	101.1	4.67	1.32	6.69	103.1	4.71	1.29	6.52

As the explanation of the cusps' extent by refraction is totally unsatisfactory, we have to assume the suggestion by RUSSELL [7], of a scattering layer composed of dust or similar medium. If the ray penetrates into this layer (Fig. 5.11) along the path SE, the planetocentric angle $2y$ replaces the refraction in the previous expressions. For this angle we have

$$y = \sqrt{\frac{2h}{a}} \qquad (5.28)$$

where h is the height of the homogeneous layer. With $q = 2.63°$, we find for it a very reasonable value $h = 2.4$ km.

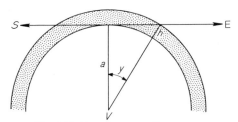

Fig. 5.11. Dust layer on Venus

That is perhaps all that the simple observations of the cusps' extent may give for our information about the state of the Venusian atmosphere. More complete information, however, may be obtained from the photometrical investigations of the cusps (5.10).

5.9. Edson's Work on Cusps Extension

EDSON contributed markedly [11] in the investigations on the cusps extension not only by collecting all available observational material but also by adding to it a new valuable photographic series obtained during the 1938 and 1940 conjunctions. From the collection of all observations since the discovery in 1790 by SCHRÖTER up to the present time, we can see only that the height of the scattering layer computed from the cusps extension is comparised between 0.5 and 12 km, the main origin of these fluctuations being the systematic errors of visual observations.

With the introduction of photographic measurements in 1938 by SLIPHER and EDSON and their continuation in 1940 by a student group at C.I.T. headed by EDSON, a new era began, bringing up some interesting results which are summed up in the following points:

a) The cusps extension is nearly neutral with a slight tendency of the growing scattering height towards the violet en of the spectrum. From the IR to the UV part of the spectrum its mean value changes from 2.9 to 3.6 km. However, reddish bright areas appear at times along the cusps extension.

b) There is a marked dissymmetry in the length and brightness of the cusps extension. This phenomenon may be connected with the large structure of the Venus atmosphere and its general circulation.

Edson also made an attempt in the theoretical interpretation of the 1938 results. He considered two models of Venus atmosphere:

i) The ordinary model with a haze layer above the opaque cloud layer. The dust-laden part of the atmosphere is illuminated by the solar rays tangent to the opaque layer. In this model the brightness of the cusps increases very rapidly towards the theoretical cusp end.

ii) The simple "cirrus deck" model, where we assume a very thin cirrus layer of a fraction of a kilometer floating in the clear atmosphere above the cloud layer. Here the increase towerds the cusps is slower than in the first case.

When compared with measurements, the second model ii) provides a better fit to the observations than the first model.

5.10. New Investigations on Cusps Extension

DOLLFUS and MAURICE [12] carried out photometric investigations of the Venus crescent during the inferior conjunction on June 20, 1964. In the principal focus of the 60 cm refractor at Pic-du-Midi they obtained in 20 days between June 6 and July 4 some 30 series of photographs. To minimise the effect of the diffused solar light in the field of the instrument, a 13 meter long mast provided at its end with a 80 cm diaphragm was fixed on the border of the dome in order to cast a shadow on the object glass. In addition to the planet images the Sun was photographed on the same plate, however, attenuated with 2 cm diaphragm and a neutral filter ($D=2$), thus enabling in this way to express the crescent intensity in solar illumination units. The images of the crescent were measured radially with the aid of the microdensimeter in different position angles counted from the point opposite to the Sun ($\alpha=0$), $\alpha=+90°$ being the north and $\alpha=-90°$ the south cusp. The final result of photometric measurements was the ratio Φ of the crescent intensity subtending the angle $1°$ to the solar illumination E_0 at the Earth's distance.

The theoretical interpretation of observed values Φ as the function of the phase angle φ can be outlined as follows. Let us divide the planetary atmosphere above the cloud level ($h=0$) in concentric layers containing $N(h)$ scattering particles in cm^3. Their scattering coefficient $R(\varphi)$ is independent of the height. Each layer contributes to the crescent intensity proportionally to the path within it $T(\varphi, \alpha, h)$ which can be computed numerically as the function of h for different values of α and φ. Each layer at the height h gives also

$$d\Phi = R(\varphi) N(h) T(\varphi, \alpha, h) dh$$

and the total intensity at the given position angle α is

$$\Phi(\varphi, \alpha) = R(\varphi) \int_{h_0}^{\infty} N(h) T(\varphi, \alpha, h) dh \qquad (5.29)$$

with the limit of the integral

$$h_0 = a \left[1 - \cos \frac{\pi - \varphi}{2} \cos \alpha \right], \qquad a = 6100 \text{ km}. \qquad (5.30)$$

At the cusps ($p = \pm 90°$) the path $T = 2\sqrt{2ah}$ is independent of the phase angle φ and the corresponding curve (Fig. 5.12) represents the scattering function of the particles. There is also a pronounced concentration of scattered light in the opposite direction of the Sun. The intensity falls to zero about at $30°$ from this direction.

In other positions angles $\alpha < 90°$ and the fixed phase angle φ the intensity of the crescent decrease as a consequence of the increasing lower limit h_0 of the integral Eq. (5.29) with the decreasing α. This decrease represents the the vertical distribution $N(h)$ of the scattering particles. In the Fig. 5.13 the observed ratio $N(h)/N_0$ is shown as the function of $1 - \cos \alpha$ for two phase angles $\varphi = 170°$ and $175°$ compared with computed curves for the following hypotheses regarding the vertical distribution $N(h)$, namely:

a) $N(h)$ varies proportionally to the air density.
b) $N(h)$ is constant up to $h = 10$ km and $N(h) = 0$ higher.
c) The same up to 7.5 km only.
d) N/h. Decreases twice as rapidly as the air density.

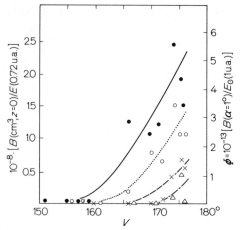

Fig. 5.12. Scattering function of dust particles (DOLLFUS and MAURICE). • $\alpha = \mp 90°$; ○ $\mp 80°$; × $\mp 60°$; △ $\mp 45°$

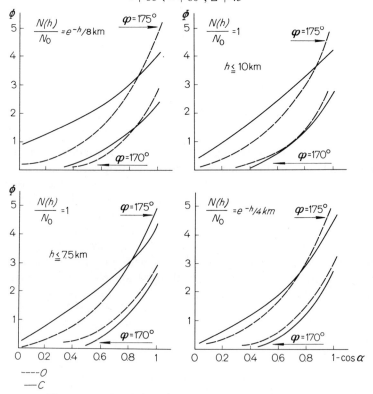

Fig. 5.13. Different models of dust distribution with height compared with observations according to DOLLFUS and MAURICE

The best agreement has been obtained with the last hypothesis (Fig. 5.13). In other words, the concentration of the scattering particles falls to $\frac{1}{2}$ with the elevation of 2.8 km.

Adopting the last hypothesis, the authors applied it to the cusps ($p=90°$) in the integral

$$\Phi(\varphi)=R(\varphi)\int_0^\infty N_0\exp\left(-2\frac{h}{H}\right)2\sqrt{2ah}\,dh \tag{5.31}$$

and its numerical value becomes $2\times 10^{20}\,N_0$ cm³.

If now d and d' are the distance of the Earth and Venus from the Sun, the crescent intensity related to the solar illumination on Venus is

$$\Phi(\varphi)\left(\frac{d}{d'}\right)^2$$

and the relative surface brightness

$$\Phi(\varphi)\left(\frac{d}{d'}\right)^2\frac{1}{(d-d')^2}=\Phi(\varphi)\times 8.8\times 10^{24}\text{ cm}^{-2}. \tag{5.32}$$

At the zero level ($h=0$) the surface brightness of 1 cm³ per unity illumination is

$$4.4\times 10^4\,\Phi(\varphi)\text{ stilb/phot.}$$

These values are given on the left vertical scale in Fig. 5.13 and their extrapolation to the phase angle $\varphi=180°$ gives for 1 cm³ and at $h=0$ 2.6 in comparison with the normal air 1.1 in 10^{-8} stilb/phot unity.

Further the variations of the intensity indicate the variations of the concentration in the ratio of about 1:5 in time and in location on the planetary limb.

Bibliography

1. LINK, F.: BAC **10**, 105 (1959) with extensive bibliography of four last transits.
2. KUIPER, G.: Ap. J. **120**, 603 (1954).
3. SHARONOW, V. V.: M. V. Lomonosov, Ak. N. S. S. S. R. Inst. Hist. IV (1960).
4. LOMONOSOV, M. V.: Works (ed. SUCHOMLINOV), vol. **5**, 113 (1902), Moskva.
5. STRUVE, O.: Sky and Telescops **13**, 118 (1954).
6. LINK, F.: BAC **1**, 77 (1949).
7. RUSSELL, H. N.: Aph. J. **9**, 284 (1899).
8. SCHÖNBERG, E.: A. N. **277**, 123 (1949).
9. LINK, F.: Mém. soc. roy. sci. Liège **18**, 148 (1957).
10. BIGG-WITHER, A. C.: Mem. roy. Astr. Soc. **47**, 97 (1883).
11. EDSON, J. B.: The twilight zone of Venus. In: Advances in Astronomy and Astrophysics (ed. KOPAL), vol. 2, p. 1. New York-London 1963.
12. DOLLFUS, A. et E. MAURICE: Compt. rend. **260**, 427 (1965).

6. Eclipse Phenomena in Radio Astronomy

6.1. Introduction

A new field in astronomy — radioastronomy — was created by the introduction of radio waves in the observational technique. Naturally we meet here with numerous analogies of the optical phenomena including the eclipses and occultations of radio sources. In these phenomena the optical refraction is replaced by the electronic refraction whose existence is conditioned by the ionised medium, for instance, the ionosphere. The radio waves traversing such a medium, i.e. the mixture: ions + free electrons + neutral particles, will be refracted according to the law of refraction with the refraction index

$$\mu^2 = 1 - \frac{N e^2}{\pi m f^2} \tag{6.1}$$

where N the number of particles per cm³
e their electrical charge and
m their mass
f the frequency of radio waves.

Due to the small mass of electrons, their refraction influence prevails over that of heavier ions. As $\mu < 1$ the sense of the electronic refraction is opposite to the optical refraction (Fig. 6.1).

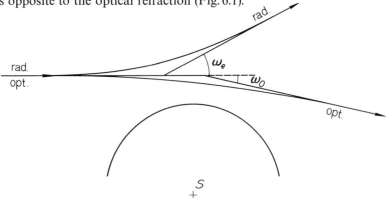

Fig. 6.1. Comparison of the optical and electronic refraction

The high sensitivity of electronic refraction to the presence of electrical particles should be noted. If we have, for instance, air of density 10^{-6}, its refraction index in the optical region will be nearly equal to unity $\mu = 1.0000000003$ and the medium behaves almost as empty space. If the same medium is only slightly ionised, i.e. every 100000th molecule having lost its electron, the refraction index for the frequency $f = 100$ MHz currently used in radio astronomy will be according to Eq. (6.1) 0.46, which is very different from unity, and producing therefore, a noticeable refraction.

In other words the electronic refraction is a very sensitive criterion of the presence of a highly rarefied atmosphere which happens to be ionised precisely because of its small density which prohibits the recombination and facilitates the penetration of exterior ionising agents.

6.2. Occultation Scene

The eclipse or occultation scene is represented in its most frequent form in Fig. 6.2. The plane I of the radio source is here at infinity. The

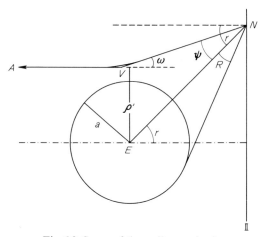

Fig. 6.2. Scene of the radio occultation

grazing ray AVN traverses the ionosphere at the minimum distance from the center

$$\rho = \frac{a + h_0}{a} \qquad (6.2)$$

and the minimum distance of its asymptote will be

$$\rho' = \mu_0 \rho. \qquad (6.3)$$

Between the geometrical distance r and the apparent distance ψ we have the relations
$$r = \psi + \omega$$
$$\psi = R\rho' \approx R\rho \qquad (6.4)$$
where R is the angular radius of the occulting body as seen from the point N which is ordinarily on the Earth.

As in the optical range, the intensity of illumination in the radio range is modified by refraction. We derive the general expression from Eq. (1.35) taking there into account the opposite sign of the refraction
$$\Phi = \left[1 + \frac{\omega}{R\rho'}\right]\left[1 + \frac{1}{R}\frac{d\omega}{d\rho'}\right]. \qquad (6.5)$$
The discussion of this formula will be given later (6.3).

6.3. Occultations of Radio Sources by the Moon

Many radio sources, among them a very bright object, can be occulted by the Moon as stated some years ago [1]. Their observations in various frequencies enable us to solve two distinct problems, namely, the determination of the exact position and size of the source, and the estimation of the upper limit for the density of the lunar atmosphere. Where the former belongs more to radio astronomy, the latter, on the contrary, is closely connected with our theme [2].

On the Moon both basic conditions for the presence of the ionosphere can be fulfilled. The solar ionisation radiation is here as intense as in the upper atmosphere on the Earth where we observe a very pronounced ionospheric region. If, on the lunar surface, the density of the atmosphere was $\rho = 10^{-9}$, it would correspond to the terrestrial ionospheric conditions at about 200 km where the electron concentration is $N = 10^5 - 10^6$ el/cm^3.

The hypothetical lunar ionosphere may be built according to Chapman's model
$$N = N_{max} \exp \frac{1}{2}\left[1 - \frac{h - h_{max}}{H} - \exp\left(-\frac{h - h_{max}}{H}\right)\right] \qquad (6.6)$$
where H is the scale height and h_{max} the level of the maximal ionisation N_{max}.

In order not to be too optimistic, we shall assume that the ionisation maximum is not yet reached even at lunar surface and this means that the part of the curve above the maximum level h_{max} is taken in consideration. In this case we can write the following approximation
$$N = N^* \exp - \frac{h}{2H} \qquad (6.7)$$

where N^* is the electron concentration on the surface. The classical formula for the refraction index may be approximated with

$$1-\mu = 4.03 \times 10^{-5} f^{-2} N. \tag{6.8}$$

For the refraction we can apply the optical formula [Eq. (4.9)] and we get

$$\omega = 4.03 \; 10^{-5} \frac{N_0}{f^2} \sqrt{\frac{\pi a_{\mathbb{C}}}{H}} = \omega^* \exp\left(-\frac{h}{2H}\right) \tag{6.9}$$

where N_0 is the electron concentration at grazing height h_0.

For the modification of the illumination by refraction we obtain with Eq. (6.5) the formula

$$\Phi = 1 - \frac{\omega}{R_{\mathbb{C}}} \left(\frac{a_{\mathbb{C}}}{2H} - \frac{1}{\rho}\right) - \frac{a_{\mathbb{C}}}{2H\rho}\left(\frac{\omega}{R_{\mathbb{C}}}\right)^2 \tag{6.10}$$

where the second term is always negative and, therefore, the increase of the illumination always takes place. Its value grows with decreasing altitude h_0 and the focussing effect takes place at the critical refraction

$$\omega_{\text{crit}} = \frac{2R_{\mathbb{C}} H}{a_{\mathbb{C}}}. \tag{6.11}$$

At the beginning or the end of the occultation $h_0 = 0$ and, therefore, the geometrical angular distance will be according to Eq. (6.4)

$$r = R_{\mathbb{C}} \frac{a_{\mathbb{C}} + h'_0}{a_{\mathbb{C}}} + \omega. \tag{6.12}$$

There is also an apparent increase of lunar radius by ω and the duration of the radio occultation is, therefore, greater than that of the optical occultation. In the central occultation, which is the most unfavourable case, by every $1''$ in the refraction the occultation is lengthened by 4 seconds of time.

6.4. Numerical Example

In Table 6.1 we have calculated ω and Φ^{-1} for exaggerated assumptions about N and H.

The course of a fictitious occultation is given in the Table 6.1 and represented in Fig. 6.3.

With the average Moon's motion of $1''$ in 2^s, the central occultation would begin about $4\frac{1}{2}$ min before the geometric beginning so that the total duration of the phenomenon would be longer by 9^m.

Table 6.1

p	ω ('')	Φ^{-1}	ψ ('')	r ('')	Δt (min)	(sec)
1.0	137	1.39	930	1067	4	34
1.1	106	1.28	1023	1129	6	38
1.2	82	1.20	1116	1198	8	56
1.3	64	1.15	1209	1273	11	26
1.4	50	1.12	1302	1352	14	04
1.5	38	1.09	1395	1433	16	46
1.6	30	1.07	1488	1518	19	36
1.7	23	1.05	1581	1604	29	40
2.0	11	1.02	1860	1872	31	14
3.0	2	1.00	2790	2792	62	04

$N^* = 10^4$ el cm^{-3}, $f = 50$ MHz, $H = 341$ km, $\omega^* = 137''$.
$\Delta t =$ the time elapsed since the geometrical occultation ($r = 930''$).

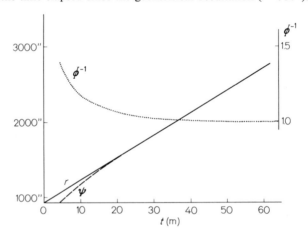

Fig. 6.3. Radio occultation by the Moon (50 MHz)

Some past observations of the occultations of the radio sources of Gemini and of the Crab nebula seem to indicate the expected refraction effect. During the occultation of the latter source on January 24, 1956 [3] the lengthening of the occultation was of 0.4^m which leads to a refraction of $13'' \pm 9''$ if we assume the refraction only on the illuminated limb of the Moon. With the scale height of 50 km, we obtain the electron density of 10^3 el/cm^3. This number says nothing about the actual density but, if we assume that 0.1% of molecules are ionised, we obtain the maximum density of the lunar atmosphere 2×10^{-13} of the terrestrial atmosphere [4]. Our method is, therefore, a thousand times more sensitive than the most refined optical method performed by DOLLFUS [5].

6.5. Occultation of Radio Source by Solar Corona

From optical measurements we know that the light of the solar corona is mainly due to the sunlight scattered on the free electrons. The decrease of the brightness with the distance from the solar border can serve with the use of an appropriate theory [6] as a means to derive the structure of the corona, i.e. the function giving the electron concentration at any height above the photosphere. The following formulae have been proposed

$$
\begin{aligned}
N &= 1.74 \times 10^7 \, \rho^{-6} \quad \text{polar regions at minimum activity [7]} \\
&= 1.74 \times 10^7 \, \rho^{-4} \quad \text{equatorial regions at minimum activity [7]} \\
&= 3.10 \times 10^7 \, \rho^{-4} \quad \text{equatorial regions at maximum activity [7]} \quad (6.13) \\
&= 1.55 \times 10^8 \, \rho^{-6} \\
&= 2.13 \times 10^8 \, \rho^{-6}
\end{aligned}
\bigg\} \text{ mean conditions } \begin{matrix} [8] \\ [9] \end{matrix}.
$$

MACHIN and SMITH [10] suggested the use of the occultation of radio sources, for instance, the Crab nebula, by the corona for an independent check of the above formulae. Without giving any mathematical theory they indicated the plausible effect of the corona i.e. the lengthening of the radio occultation in comparison with the optical phenomenon. However, the mathematical theory was given afterwards [6].

The starting formula of the theory will be

$$N = a\rho^{-n} \qquad (6.14)$$

giving the electronic structure of the corona which is supposed to have a spherical symmetry. With the aid of Eq. (6.14), we find the refraction index

$$\mu^2 = 1 - ac\rho^{-n} = 1 - 8.06 \times 10^{-5} f^{-2} N. \qquad (6.15)$$

The classical form of the refraction integral

$$\omega = 2 \int_{\rho_0}^{\infty} \frac{d\mu}{\mu} \, \text{tg}\, i \qquad (6.16)$$

together with the theorem of the invariance

$$\mu \rho \sin i = \mu_0 \rho_0 \qquad (6.17)$$

will lead to the expression

$$\omega = acn\rho_0 \frac{\mu_0}{\mu^2} \int_{\rho_0}^{\infty} \frac{d\rho}{\rho^{n+1} \sqrt{\rho^2 - \rho_0^2}} \frac{1}{\sqrt{1 - ac\dfrac{\rho^{-n+2} - \rho_0^{-n+2}}{\rho^2 - \rho_0^2}}} \qquad (6.18)$$

which may be simplified considering that the maximum of the term

$$x = a\,c\,\frac{\rho^{-n+2} - \rho_0^{-n+2}}{\rho^2 - \rho_0^2} \qquad (6.19)$$

will be attained for $\rho = \rho_0$ and its value will be

$$x_0 = \tfrac{1}{2}a\,c(n-2)\,\rho_0^{-n} \leq \tfrac{1}{50} \qquad \text{for } \rho_0 > 3 \qquad (6.20)$$

i.e. practically negligible. We have then approximately

$$\omega = a\,c\,n\,\rho_0 \int_{\rho_0}^{\infty} \frac{d\rho}{\rho^{n+1}\sqrt{\rho^2 - \rho_0^2}} = B f^{-2} N_0 \qquad (6.21)$$

where the factor B depends on the exponent n as is shown in the following table.

Table 6.2

n	1	2	3	4	5	6
B	8.06	12.66	16.12	18.99	21.49	23.74

The value of the integral varies with n which may change to some extent within the integration interval. In order to see this influence more clearly let us introduce a new variable

$$Z = \sqrt{\rho^2 - \rho_0^2} \qquad (6.22)$$

which gives with equation

$$I = \int_0^{\infty} \frac{dZ}{\rho^{n+2}}. \qquad (6.23)$$

Thus growing distances ρ enter in to this integral with rapidly decreasing weights and it is sufficient to adopt the right form of the electron concentration at the lower limit of the integral. This is, besides, the general characteristic of all ray deviations discussed in this book.

6.6. Modification of Light Intensity

By the analogy with Eq. (6.4) we have for the relation between the geometrical r and apparent distance ψ

$$r = \psi + \omega = \rho R + \omega$$

and for the modification of the light intensity

$$\Phi = 1 - (n-1)\frac{\omega}{\rho R} - n\left(\frac{\omega}{\rho R}\right)^2. \qquad (6.24)$$

The discussion leads to the following conclusions: Far from the Sun where $\omega=0$, we have $r=\psi$ and $\Phi=1$. With decreasing distance from the Sun, φ decreases too, i.e. the intensity of the illumination of the terrestrial plane increases more and more rapidly (Fig. 6.4) and the distance

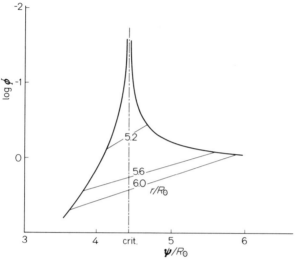

Fig. 6.4. Computed light curve of the radio occultation by solar corona (38 MHz)

ρ_{crit}, called the critical distance, where $\Phi=0$, the intensity grows to infinity. The condition for it is according to Eq. (6.24)

$$\frac{\omega_{\text{crit}}}{\rho_{\text{crit}} R} = \frac{1}{n} \tag{6.25}$$

or with Eq. (6.21) the critical distance will be

$$\rho_{\text{crit}} = \left(\frac{a B_n}{R f^2}\right)^{\frac{1}{1+n}}. \tag{6.26}$$

Having passed the critical distance the intensity decreases. Its negative value has only a geometrical significance.

Let us follow the course of the geometrical distance with decreasing distance of rays (Fig. 6.5). The conditions of the minimum are

$$\frac{dr}{d\rho_0} = R - n a B \rho^{-n-1} f^{-2} = 0$$

$$\frac{d^2 r}{d\rho_0^2} = n(n+1) a B \rho^{-n-2} f^{-2} > 0. \tag{6.27}$$

The solution gives

$$\rho = \rho_{\text{crit}}.$$

The minimum of r arrives, therefore, at the critical distance where

$$r_{crit} = \frac{n+1}{n} R \rho_{crit}$$

and (6.28)

$$r_{crit} = \frac{n+1}{n} R \left(\frac{a B_n}{R f^2}\right)^{\frac{1}{1+n}}.$$

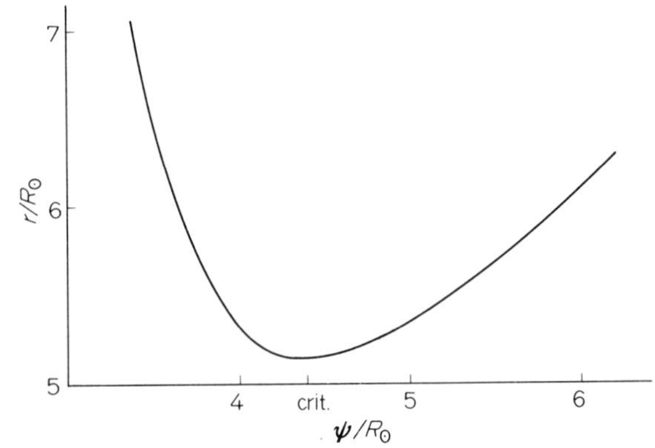

Fig. 6.5. Relation between the geometric (r) and apparent (ψ) angular distance during the occultation by solar corona

There is also the radio limb of the Sun, which is much further then the optical limb as it is shown for some plausible solar situations in the following table.

Table 6.3

Coronal structure	38 MHz	81.5 MHz	210 MHz
Polar region at minimum	3.23	2.60	
Equatorial region at minimum	2.96	2.38	
Equatorial region at maximum	3.19	2.57	
Middle conditions	4.42	3.54	2.70

6.7. Observations of Coronal Occultations

The observations of Crab nebula, a know radio source, during its close approach to the Sun which takes place every year by the middle of June may be a good verification of the refraction theory. At the minimum distance $4.6 R_\odot$, the photometrical effect according to Table 6.3 and Fig. 6.4 should be appreciable in the 38 MHz band.

However, the first observations by MACHIN and SMITH [11] as well as all subsequent observations by HEWISH [12] and others showed instead of the expected amplification of the intensity a decline beginning at several solar radii (Fig. 6.6) and nearly independent of the frequency used.

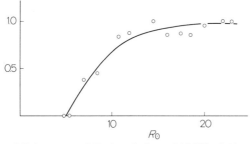

Fig. 6.6. Observed light curve of Crab nebula at 81 MHz during the occultation by solar corona

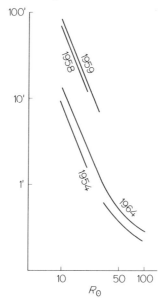

Fig. 6.7. The angular scattering at 38 MHz by solar corona

Therefore, HEWISH [13] proposed the scattering of radio waves on the irregularities of the coronal structure as the main origin of the observed decline. The filamentary structure as the extension of coronal rays or the density fluctuations generated by the plasma instability should be taken into consideration. Besides this phenomenon, the discovery of interplanetary scintillation of radio sources has proved the extension of the plasma irregularities beyond the Earth's orbit.

The scattering effect of solar corona at a given distance depends on the phase of solar activity (Fig. 6.7). The angular scattering was much higher in 1958/59, period of solar maximum, than in 1954 or 1964, years of solar minima. Any further discussion of this matter lies in fact beyond the scope of this book.

6.8. Occultation of the Mariner-IV Space Probe by Mars

"Approximately 1 hour after its closest approach to Mars on July 15, 1965, the Mariner-IV spacecraft disappeared beyond the limb of the planet, as seen from Earth, and remained in occultation for approximately 54 minutes." With these sober words KLIORE et al. introduce the report [14] one perhaps the most interesting achievement in the sphere of eclipse phenomena produced artificially by human skill and intelligence.

Without enterning fully into the matter of a very complicated electronic instrumentation enabling the detection and measurements of small changes of phase, we shall content ourselves with an outline of its general principles. The emitted frequency (2300 MHz) derived from a rubidium vapour standard was sent to the spacecraft through a 26-metre parabolic antenna. On the spacecraft a transponder coherently retransmitted the received signal to the Earth after the shift of the frequency due to several causes among them the Mars atmosphere. This signal captured at the ground station is compared in the receiver with the same Rb-standard. It may be, of course, assumed that its drift during the round trip of the signal to the probe and back (0.5 hour) is negligible. The comparison of frequencies is made by means of frequency operations ($\times$, :, $+$) with the reference frequency in order to obtain finally the pure Doppler frequency which is cumulated and counted in 1-second intervals. These values were compared with those predicted by the orbital motion and the resulting differences are assumed to be an effect of the Mars atmosphere.

We wish to emphasize that a very ligh level of accuracy is necessary in both frequency measurements and in orbital motion of the probe. The phase change due to the Mars atmosphere amounts to about 30 cycles or wave lengths and that due to the ionosphere only about 10 cycles. These changes are the result of the subtraction from the total frequency phase change of about 3×10^{11} cycles produced by the probe motion, reference phase change, motion of the ground station (3×10^7) and others.

The data of Doppler tracking taken just before and after the planetary approach give the trajectory of the probe during the occultation with a very high accuracy, the range rate of the probe being known at ± 0.0015 m/sec.

This system of operation, called a two-way or coherent mode, was supplemented by a one-way operation during the first 9 seconds after the emersion when no transmission from the Earth had been received by the probe. During this interval of time the probe transmitter was controlled by an onboard crystal oscillator of smaller stability than the Rb-standard.

The data about the grazing areas on the surface of Mars at the entry and exit are given in the Table 6.4. Mars occulted the probe with the sunlit limb during a winter afternoon at 50° S latitude. The exit took place in summer on the night limb before sunrise at 60° N latitude. We must bear in mind these very different conditions when discussing the results of ionospheric and atmospheric data.

The general feature of the residual phase change as the function of time derived from one of three tracking stations is shown for the entry

Table 6.4

	Entry	Exit
Latitude	50.5° S	60° N
Longitude	177° E	34° W
Location	Elektris Mare Chronium	Mare Acidalium
Solar zenith angle	67°	104°
Local time, on the surface of Mars	13:00	00:30
Surface refractivity, $(\mu_0 - 1) \times 10^6$	3.6±0.2	4.2±0.3
Scale height near surface, km	9±1	12±1
Surface number density, 10^{17} mol/cm^3 100% CO_2 80% CO_2, 20% A	 1.9±0.1 2.1±0.1	 2.25±0.15 2.45±0.15
Surface mass density, 10^{-5} g/cm^3 100% CO_2 80% CO_2, 20% A	 1.43±0.1 1.55±0.1	 1.65±0.15 1.85±0.15
Surface temperature, °K 100% CO_2 80% CO_2, 20% A	 180±20 175±20	 240±20 235±20
Surface pressure, mb 100% CO_2 80% CO_2, 20% A	 4.9±0.8 5.2±0.8	 8.4±1.3 8.8±1.3
Maximum electron density, ionosphere, el/cm^3	$9.0±1.0 \times 10^4$	4×10^3
Altitude of maximum, km	123±3	
Electron scale height above maximum, km	22±3	—

in the Fig. 6.8. The curve displays in its first part the influence of the ionosphere with the maximum effect at 02 30 10 and the final rise due to the neutral low atmosphere.

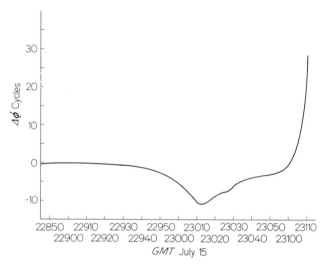

Fig. 6.8. Phase change along the ray in the Martian atmosphere (Mariner IV)

The fundamental equation for the phase path change along the ray is given by

$$\Phi = \frac{1}{\lambda} \int_{-\infty}^{+\infty} (\mu - 1)\, ds. \tag{6.29}$$

In the low atmosphere this equation may be used if we choose a model of the atmosphere which enables us to compare the computed and observed values of Φ. This model should be characterized by the surface refractivity $\mu_0 - 1$ and the scale height H defining the exponential structure of the atmosphere

$$\mu - 1 = (\mu_0 - 1) \exp\left(-\frac{h}{H}\right). \tag{6.30}$$

If μ and H have been found to be the best fitting values to the observed curve (Fig. 6.8), we can proceed to the estimation of the atmospheric composition and subsequently the surface temperature. The authors adopted two models: one with 100% content of CO_2 and another with 80% of CO_2 and 20% of Ar (Argon). The corresponding values of atmospheric parameters are given in Table 6.4.

The inspection of the numbers given above deserves some comment. First the surface temperature at the exit region is markedly higher than at the entry region. This must be understood as a seasonal effect, the former region having a late summer in contrast with the latter region where it was the end of the winter at the time of Mariner-IV fly-by.

Most intriguing, however, is the pressure difference or better the pressure ratio between the two Martian spots. One might speculate about the possible level difference of both regions; it is interesting to note that the pressure difference correspond to the level difference of 5.8 km for a scale height $H = 11$ km. And precisely the planetary radii determined from the occultation moments lead to the difference of 5 km between both regions and in the right direction. Unfortunately the standard error is rather large ± 5 km. Otherwise one might be tempted to detect here the self-consistency of both results although, however, the flattening is imperfectly known. No clear conclusion has been drawn by the authors regarding the cause of the observed pressure difference between the two regions.

In the ionospheric part of the path curve the refraction index is to be taken according to Eq. (6.8) in which the electron concentration N may be adopted according to the Chapman's model [Eq. (6.6)].

From this part of the phase path curve the best fitting model has been derived giving the electron distribution in height. Its parameters are given at the bottom of Table 6.4. A preliminary discussion of the exit data failed to show any detectable ionosphere of the electron concentration higher than 1/20 of the entry value. This is, of course, the natural consequence of night conditions in contrast with day conditions at the entry. Adopting the Chapman layer model which satisfies well the observations above the peak concentration, we may compute the electron concentration 1.5×10^5 el/cm^3 and the total electron content in the vertical column 7×10^{11} el/cm^2. Corresponding terrestrial values at $Z_\odot = 0°$ are 10^6 el/cm^3 and 10^{13} el/cm^2 with important variations with different factors such as the season, the latitude and solar activity.

The near or complete absence of a magnetic field on Mars as it results from another Mariner-IV experiment [15] facilitates the theoretical study of the Mars ionosphere in comparison with the analogous task in the terrestrial atmosphere. The Earth's magnetic field, controlling the motion of solar particles and affecting the ionospheric motions, renders difficult the understanding of many ionospheric phenomena. Results of studies of the Martian ionosphere may also help in separating and understanding some complicated phenomena in the terrestrial ionosphere.

Summing up, we may conclude that the Martian atmosphere is thinner than was previously thought according to visual observations. We may further hope that a more refined analysis of additional data as well as a more complete theory will bring some corrections or ameliorations to the results given above.

6.9. Further Implications of the Mariner-IV Mission

The new Mariner-IV results correct some erroneous views expressed earlier in comparing the Martian and terrestrial atmospheres. Since on Mars the gravity is about 62% lower, the density decrease with height would be *ceteris paribus* much slower than on the Earth. Thus, in spite of the lower density near the surface of Mars, it would become larger at about 30–50 km than the density at the same level in the Earth's atmosphere and this prevalence would be maintained until the upper atmosphere.

The above given Mariner-IV results radically changed this conception. That the atmosphere is composed mainly of CO_2 has the following reasons: the weak CO_2 absorption lines near 8700 Å indicate the equivalent path along the vertical direction of about 55 m of gas reduced to standard temperature and pressure. This amount of CO_2 would contribute by 4 mb to the surface pressure. As Mariner-IV data give about 5 mb for the total pressure little space is left for other gases and, therefore, the main constituent of the Martian atmosphere should be carbon dioxide.

If now the atmosphere is mainly composed of heavy CO_2 ($m=44$) and is quite cold ($T=180°$), the scale height according to Eq. (4.32) will not be very different from the terrestrial value, as the smaller gravity g is nearly compensated by low temperature T and high molecular mass m. We can, therefore, say that roughly the density ratio on both planets does not change greatly with altitude.

The lower density of the upper atmosphere has also a practical implication for future Martian projects aiming to place an artifical satellite in orbit around the planet. Previous dense atmospheric models required an orbital height of several thousand kilometers in order to guarantee a lifetime of about 50 years. The new Mariner-IV model of the atmosphere demands only some hundred kilometers which greatly simplifies the surface observing devices.

If CO_2 is the principal atmospheric constituent, we may expect that at a 70–80 km level this gas is dissociated by solar ultraviolet radiation

$$CO_2 + \varepsilon = CO + O$$

and the resulting component O should be distributed by the diffusion process and the diffusive separation will exist. Thus the atomic oxygen should prevail over other gases in the upper atmosphere. As the Martian atmosphere is very cold, 180° K near the surface and 80° K in the ionosphere, there is a great chance of meeting in this hight interval the frozen particles of CO_2.

The above semiquantitative considerations may be supplemented by a more elaborate theory in order to explain the observed facts and in addition enable us to compute the detailed structure of the Martian atmosphere. This subject is, however, beyond the scope of this book.

6.10. Occultation of the Mariner-V Space Probe by Venus

Similar to the Mariner-IV mission towards Mars was that of Mariner-V towards Venus on October 19, 1967, when this probe passed within about 4100 km of the surface of Venus [16]. Observed from the Earth, however, the occultation took place.

Fearing a big attenuation effect by refraction, it was decided against the use of the two-way mode and the much surer one-way mode was adopted. Furthermore the high gain antenna of the probe was not directed exactly towards the Earth in order to avoid this harmful refraction influence.

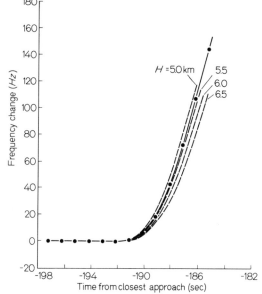

Fig. 6.9. Frequency change at the entry (night) observed by Mariner V (full line) compared with different atmospheric models (dashed line)

For the interpretation of frequency change measurements, several models of exponential atmosphere with the scale height lying between $5.0 < H < 6.5$ were adopted and compared with observations (Fig. 6.9). The entry (night) data seem to give the best fit with $H = 5.4$ km in the height range 10 km of the neutral upper atmosphere. The same may be valid for the exit (day) data with some perturbations due to the very intense day ionosphere (Fig. 6.10).

16 Link, Eclipse Phenomena

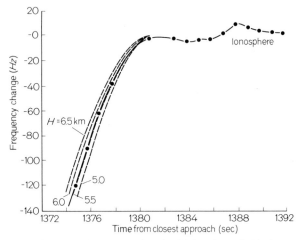

Fig. 6.10. The same as in Fig. 6.9 but for exit (day)

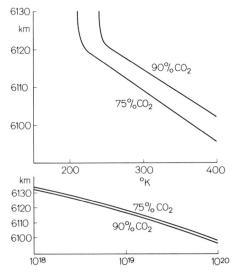

Fig. 6.11. Temperature (above) and density (below) curve of the night atmosphere of Venus (Mariner V)

The above value of H gives only the ratio of $T(m)$ Eq. (4.32). If on assumes from other evidence for the temperature $T = 230°$ K, the value $H = 5.4 \pm 0.2$ km gives for the molecular weight $m = 39.0$ to 42.0 corresponding to carbon dioxide percentage of about 69 to 87% provided that the other component is nitrogen ($m = 28$); pure CO_2 would give $m = 44$. With this assumption of constant composition, we may compute the temperature curve in the lower part of the atmosphere (Fig. 6.11).

The refractivity $c\rho$ has been computed from the frequency change (Fig. 6.11) and it may be converted into the densities ρ if one adopts the above composition. The density curve leads at the bottom of the observable atmosphere, at the distance of 6095 km from the centre, to the value of 10^{20} molecules/cm^3 in comparison with the terrestrial value 2.7×10^{19}. The corresponding pressure with the temperature adopted from Fig. 6.11 will be about 5 atm.

The ionospheric part of the occultation curve gave on the day side (exit) the electron profile with a maximum of 5.5×10 el/cm^3 at 6190 km and some indication of a secondary maximum at 6175 km.

One interesting implication follows from the structure of the lower part of the neutral atmosphere. Adopting there the most probable value of $H = 11$ km and nearly 90% of CO_2 giving for the normal refractivity $c = 4 \times 10^{-4}$, the critical density [Eq. (4.19)]

$$\rho > \frac{1}{ac\beta} = \rho^* = 4.5, \tag{6.31}$$

which is necessary to imprison the inclined rays (4.3) could be attained a little below the still measured level at 6095 km. Let us remember that the radar determinations [17] gave the surface radius 6056 ± 1.5 km. As the visual radius of Venus is nearly 6100 km, the capture level may be identical with the visibility level for horizontal rays, if we admit some small corrections in the above relation lying within the limits of the errors of measurements.

Bibliography

1. LINK, F., and L. NEUŽIL: BAC **5**, 112 (1954).
2. LINK, F.: BAC **7**, 1 (1965); Radio Astronomy Symp. Jodrell Bank 1955.
3. COSTAIN, C. H., B. ELSMORE, and G. R. WHITFIELD: M. N. **116**, 380 (1956).
4. ELSMORE, B.: IAU Symp. (Radio Astr.) No 9 (1959).
5. DOLLFUS, A.: Compt. rend. **234**, 2046 (1952).
6. LINK, F.: BAC **3**, 6 (1952).
7. VAN DE HULST, H. C.: BAN **11**, 135, 150 (1950).
8. ALLEN, C. W.: MN. **107**, 426 (1947).
9. BOGORODSKIJ, A. F., i N. A. CHINKULOVA: Publ. Kiev **4**, 3 (1905).
10. MACHIN, K. E., and F. G. SMITH: Nature **168**, 599 (1951).
11. MACHIN, K. E., and F. G. SMITH: Nature **170**, 319 (1952).
12. HEWISH, A.: IAU Symp. Radio Astr. No. 9, 268 (1959).
13. HEWISH, A.: Proc. Roy. Soc. (London) A **228**, 238 (1955).
14. KLIORE, A., D. L. CAIN, G. S. LEVY, V. R. ESHLEMAN, G. FJELDBO, and F. D. DRAKE: Science **149**, 1243 (1964).
15. O'GALLAGHER J. J., and J. A. SIMPSON: Science **149**, 1233 (1964).
16. KLIORE, A., G. S. LEVY, G. FJELDBO, and S. I. RASOOL: Science **158**, 1683 (1967).
17. ASH, M. E., I. I. SHAPIRO, and W. B. SMITH: Astr. J. **72**, 338 (1967).

7. Einstein's Deflection of Light

Einstein's deflection of light may be considered as a particular case of the refraction field in the neighbourhood of a material body with all its consequences such as the formation of images and the modification of light intensity. The lunar eclipse theory was, therefore, for us the starting point of a general investigation of this problem [1]. It was also the first correct treatment of the photometrical side and the most general examination of gravitational imagery. The subsequent work by other authors was mainly partial or cursory and, therefore, our theory will serve as the frame-work of this chapter. Of course, we have no intention of overlooking these scientists, as shall compare their results with ours and give prominence in the discussion to their applications in stellar astronomy.

7.1. Einstein's Deflection of Light

Before approaching the main problem we must clarify the theory of the single ray deflection. Let us designate the mass m of the deflecting body and a its radius. The equation of the light path will be [2] in polar coordinates (Fig. 7.1)

$$R = r \cos \varphi + \frac{m}{R}(r \cos^2 \varphi + 2r \sin^2 \varphi) \tag{7.1}$$

or in orthogonal coordinates

$$x = R - \frac{m}{R} \frac{x^2 + 2y^2}{\sqrt{x^2 + y^2}} \tag{7.2}$$

with

$$R = \overline{DV}.$$

The light path has two asymptotes for $y \to \infty$

$$x = R \pm \frac{2m}{R} y \tag{7.3}$$

and their angle at C

$$\omega = \frac{4m}{R} \tag{7.4}$$

give Einstein's deflection of light by the body at C.

Einstein's Deflection of Light

For the Sun we have with

$$m = 1.47 \text{ km} \quad \text{and} \quad a = 6.95 \times 10^5 \text{ km}$$

the grazing deflection

$$k = 1.75'' = \omega_0.$$

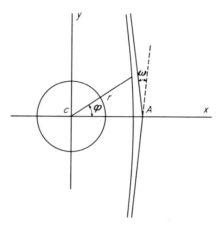

Fig. 7.1. Einstein's deflection of light

The partial deflection between the two points φ_1, r_1 and φ_2, r_2 will be

$$\omega_{1,2} = \frac{m}{R} \int_{\varphi_1}^{\varphi_2} (\cos \varphi + 3 \sin^2 \varphi \cos \varphi) \, d\varphi \tag{7.5}$$

which gives for

$$\varphi_1 = -\frac{\pi}{2} \quad \text{and} \quad \varphi_2 = +\frac{\pi}{2}$$

again the Eq. (7.4).

In addition it is perhaps useful to know the minimal distance of the asymptote from the center C

$$R'_0 = R\left(1 - \frac{8m^2}{R^2}\right) \tag{7.6}$$

and the minimal distance of the trajectory from the same point

$$R_0 = R\left(1 - \frac{m}{R}\right). \tag{7.7}$$

In most cases we have, therefore, with a high degree of approximation

$$R = R'_0 = R_0. \tag{7.8}$$

In subsequent pages we chose for the unity of the mass the solar mass $m_\odot$ and for the distance the solar radius $a_\odot$. We have, hence,

$$\omega = K k \frac{a}{R} \qquad (7.9\,\text{a})$$

with

$$K = \frac{m}{a} \frac{a_\odot}{m_\odot}. \qquad (7.9\,\text{b})$$

The informational values of K for some celestial objects are given in the Table 7.1.

Table 7.1

	Sun	Red giant	Red dwarf	White dwarf	Galaxy
M	1	10	$\frac{1}{2}$	1	10^{10}
a	1	100	1	0.01	2×10^{10}
K	1	0.1	0.5	100	0.5
kK	1.75″	0.175″	0.87″	175″	0.87″

7.2. Dioptrics of Einstein's Deflection

Let us adopt the same scene as for lunar eclipses with the illuminating plane I containing the point source M, the deflecting body D and the illuminated plane II with the observing point N (Fig. 7.2).

The position of point M is defined by the angle

$$r = (\alpha_1 + \alpha_2) \frac{R}{a} - \omega \qquad (7.10)$$

which gives with Eq. (7.9 a) quadratic equation for R

$$(\alpha_1 + \alpha_2) R^2 - a r R - K k a^2 = 0 \qquad (7.11)$$

with the solution

$$R_{1,2} = a \frac{r \pm \sqrt{r^2 + 4(\alpha_1 + \alpha_2) K k}}{2(\alpha_1 + \alpha_2)}. \qquad (7.12)$$

In other words the observer at N receives two conjugate rays, 1 and 2, one passing at the minimal distance R_1 from D and the other at the distance R_2. The latter has a negative sign which corresponds to the opposite position of ray 2 in respect to ray 1. The observer at N sees in consequence instead of one point in the geometrical direction

$$\measuredangle MND \equiv g = \frac{\alpha_2}{\alpha_1 + \alpha_2} r \qquad (7.13)$$

two images M_1 and M_2 in the apparent conjugate directions

$$\rho_{1,2} = \frac{R_{1,2}}{l_2} = \alpha_2 \frac{r \pm \sqrt{r^2 + 4(\alpha_1 + \alpha_2) K k}}{2(\alpha_1 + \alpha_2)}. \tag{7.14}$$

For the same reason as before ($R_2 < 0$) we have again $\rho_2 < 0$.

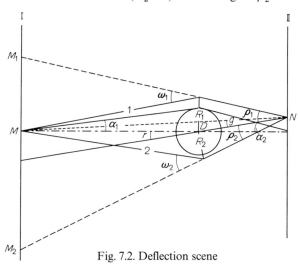

Fig. 7.2. Deflection scene

We shall now discuss the general behaviour of the two images $M_{1,2}$ in the function of the decreasing angle r, that is when the light source M is approaching from infinity to the axis (Fig. 7.3).

Far from the axis, i.e. for the high values of r when $r^2 \gg 4(\alpha_1 + \alpha_2) K k$, we have simply

$$\rho_1 = \frac{\alpha_2}{\alpha_1 + \alpha_2} r = g \tag{7.15}$$

and $\rho_2 = 0$.

The first image M_1 is, therefore, practically in coincidence with M and the second image M_2 with D.

When N is approaching the axis, the first image M_1 separates itself from M and remains retarded behind it. The second image M_2 moves away from D in the opposite direction (Fig. 7.3).

When

$$r \to 0 \quad \text{then} \quad \rho_{1,2} \to \rho_* = \alpha_2 \sqrt{\frac{K k}{\alpha_1 + \alpha_2}}$$

$$= 9.026'' \times 10^{-2} \sqrt{\frac{m}{m_\odot} \frac{l_1}{l_2} \frac{1}{l_2 \,(\text{psc})}} \tag{7.16}$$

the two images $M_{1,2}$ are both approaching the same critical circle ρ_*, M_1 from its exterior and M_2 from its interior. When $r=0$ both images are extended to the critical circle and both rays 1 and 2 pass at the critical distance

$$R_* = a \sqrt{\frac{Kk}{\alpha_1 + \alpha_2}}. \tag{7.17}$$

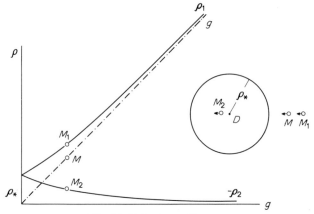

Fig. 7.3. Behaviour of images

7.3. Photometry of Einstein's Deflection

To calculate the illumination at N from the point source at M, we need only to use our Eq. (1.35) from lunar eclipses

$$\frac{i}{I} = \Phi = \left[1 - \frac{\omega}{\alpha_1 + \alpha_2} \frac{a}{R}\right] \left[1 - \frac{d\omega}{dR} \frac{a}{\alpha_1 + \alpha_2}\right] \tag{7.18}$$

where we put for ω the value from Eq. (7.9a) and for

$$\frac{d\omega}{dR} = -Kk \frac{a}{R^2}. \tag{7.19}$$

Thus

$$\Phi = 1 - \frac{K^2 k^2}{(\alpha_1 + \alpha_2)^2} \frac{a^4}{R^4} \tag{7.20}$$

or

$$\frac{I}{i} = \Phi^{-1}. \tag{7.21}$$

There I is the modified intensity of the illumination and $i=1$ its normal value in the absence of the deflecting body.

We shall now follow the variation of s with decreasing distance R. At grat distances Φ is very slightly smaller than unity and in consequence we have only an insignificant increase of the illumination (Fig. 7.4). With the decreasing distance R the value of I grows and attains infinity at

$$R_* = a \sqrt{\frac{Kk}{\alpha_1 + \alpha_2}} \tag{7.17}$$

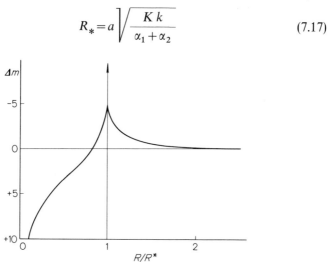

Fig. 7.4. Light curve of Einstein's occultation

which corresponds to the critical circle ρ_* [Eq. (7.16)]. The rays passing the deflecting body at this distance are deflected to the point at the axis $r = 0$.

For the ray passing nearer than R_*, the value of I becomes negative. The meaning of it is only geometric as for the sign of r. At the distance

$$R = 2^{-\frac{1}{4}} R_* = 0.84 R_* \tag{7.22}$$

$\Phi = -1$ and there is no modification of the intensity. Still nearer to the deflecting body, we have the attenuation of light which continues up to the surface of the body.

It depends on the actual size (a) of the deflecting body whether all the phases discussed above would be observable.

The limiting case $\bar{r} \to 0$, though mathematically based, has no physical meaning and is entirely unrealistic for two reasons. Firstly, we never have a perfect point source. Even the smallest disk of apparent radius g_1 gives a finite value of the illumination, as will be shown later [Eq. (7.45)]. Secondly, we never have a point receptor of light. This is generally an object glass or mirror of finite aperture $2s$. Hence the total flux received by it should be considered. If the optical axis of the object glass coïncides with the direction $r=0$ ($g=0$) we shall compute the flux captured by it (Fig. 7.5). The concentric ring ds has the area expressed by

$$ds = 2\pi r \, l_2^2 \, dr \tag{7.23}$$

and the flux received according to Eq. (7.35) will be for $r \to 0$

$$dF = I \, ds = 2\pi \, l_2^2 \sqrt{(\alpha_1 + \alpha_2) K k} \, dr. \tag{7.24}$$

Then the total flux intercepted by the object glass has a finite value

$$F = 2\pi \, l_2 \, s \sqrt{(\alpha_1 + \alpha_2) K k}. \tag{7.25}$$

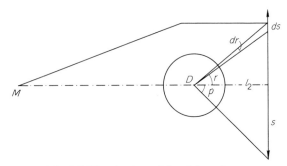

Fig. 7.5. Illumination of the object glass

In the absence of the deflecting body it would be

$$F_0 = \pi s^2 \tag{7.26}$$

and the ratio of both fluxes is

$$\frac{F}{F_0} = \frac{\sqrt{4(\alpha_1 + \alpha_2) K k}}{p} \tag{7.27}$$

where p is the apparent size of the radius s viewed from the deflecting body.

7.4. Other Expression of Illumination

We can derive another independent expression for the illumination by the light point source. The solar element (Fig. 7.6) at M

$$dq = x \, d\varphi \, dx \tag{7.28}$$

appears from N in two different directions ρ_1 and ρ_2. Let us consider the first direction where we see the element

$$dq_1 = x_1 \, d\varphi \, dx \tag{7.29}$$

Other Expression of Illumination

and according to Fig. 7.6 we have

$$x = r\,l_1$$
$$dx = l_1\,dr$$
$$x_1 = (l_1 + l_2)\,\rho_1 \tag{7.30}$$
$$dx_1 = \frac{a}{2\alpha_1}\left(1 + \frac{r}{\sqrt{r^2 + 4(\alpha_1 + \alpha_2)\,Kk}}\right).$$

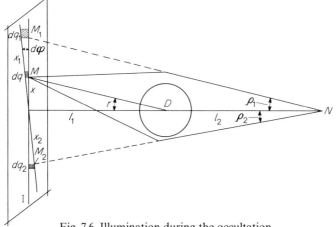

Fig. 7.6. Illumination during the occultation

We have, therefore, for the two elements

$$dq = \frac{a^2}{\alpha_1^2} r\,dr\,d\varphi$$
$$dq_1 = \frac{a^2}{4\alpha_1^2}\left(r + \sqrt{r^2 + 4(\alpha_1 + \alpha_2)\,Kk}\right)\left(1 + \frac{r}{\sqrt{r^2 + 4(\alpha_1 + \alpha_2)\,Kk}}\right)dr\,d\varphi. \tag{7.31}$$

The illumination at N in the absence of the deflecting body will be

$$i = \frac{b\,dq}{(l_1 + l_2)^2} \tag{7.32}$$

and with the deflecting body

$$I = \frac{b\,dq_1}{(l_1 + l_2)^2} \tag{7.33}$$

where b is the luminance of the stellar surface unchanged by any dioptric or catoptric process. Finally, we obtain for the two directions ρ_1 and ρ_2

the expression

$$\frac{I_{1,2}}{i} = \frac{r \pm \sqrt{r^2 + 4(\alpha_1 + \alpha_2)Kk}}{4r\sqrt{r^2 + 4(\alpha_1 + \alpha_2)Kk}}. \tag{7.34}$$

The total illumination by M at N is the difference $I_1 - I_2$, while I_2 is negative for geometrical reasons (7.2)

$$I_t = \frac{r^2 + 2(\alpha_1 + \alpha_2)Kk}{r\sqrt{r^2 + 4(\alpha_1 + \alpha_2)Kk}} \cdot i = \frac{g^2 + 2\rho_*^2}{g\sqrt{g^2 + 4\rho_*^2}} \cdot i. \tag{7.35}$$

On the contrary, the sum

$$I_1 + I_2 = i. \tag{7.36}$$

The equations for illuminations [Eqs. (7.34)–(7.35)] are symmetrical with respect to the distances l_1 and l_2. There exists, therefore, between the planes I and II dioptrical and photometrical reciprocity in analogy with lunar eclipses (1.2.10).

There exists a difference in the path between the two directions ρ_1 and ρ_2 as was first shown by REFSDAL [3]. Neglecting the very small influence of the curved parts on the light paths we find for the path difference on the straight parts (Fig. 7.7)

$$\Delta s = l_1(\cos\tau_1 - \cos\tau_2) + l_2(\cos\rho_1 - \cos\rho_2). \tag{7.37}$$

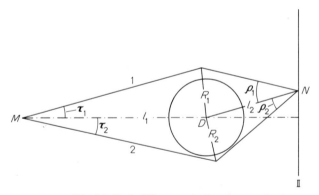

Fig. 7.7. Path difference during the occultation

Noting that

$$\tau_{1,2} = \frac{R_{1,2}}{l_1}, \quad \rho_{1,2} = \frac{R_{1,2}}{l_2}$$

and with the aid of Eqs. (7.12) and (7.13)

$$\Delta s = \frac{l_2}{2} \frac{l_1 + l_2}{l_1} g\sqrt{g^2 + 4\rho_*^2}. \tag{7.38}$$

7.5. Illumination by Stellar Disk

For the computation we adopt the scene as for lunar eclipses, and the scheme for the theoretical treatment will be also the same. An elementary ring m on the occulted star described by the cones of two conjugate rays (Fig. 7.8) illuminates the point N. If di is the mean light

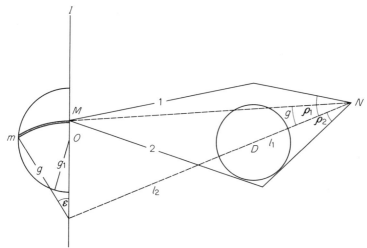

Fig. 7.8. Illumination by stellar disk

intensity of this ring and I the modification of the intensity, the elementary illumination at N will be

$$dE = I\,di \qquad (7.39\,\text{a})$$

and the global illumination is given by the integral extended over the whole disk

$$E = \int I\,di \qquad (7.39\,\text{b})$$

For the mean light intensity di we can use, after an appropriate change of scale, our Table 1.6 of lunar eclipses. As this table is computed for solar disk of 16' radius all arguments must be multiplied there by the factor $g_1/16'$.

The evaluation of the integral [Eq. (7.39 b)] is made by numerical or graphical methods. Care must be taken to verify whether the second ray in the direction ρ_2 can attain the point N. As long as $R_2 > a$, the Eq. (7.34) is to be used. In the case where $R_2 < a$ only the first component I_1 can illuminate the point N and consequently the value of I_1 is to be used in the integration.

During the central occultation we have for the illumination by the ring between r and $r+dr$

$$dE = 2\pi g \, dg \, b \frac{I}{i} \tag{7.40}$$

where the surface brightness b is expressed by the conventional formula [Eq. (1.16)].

The modification of the intensity at small distances $g < \frac{\rho_*}{5}$ is with high degree of approximation

$$\frac{I}{i} = \frac{\rho_*}{g}. \tag{7.41}$$

Thus by introducing [Eq. (1.16)] and [Eq. (7.41)] into [Eq. (7.40)] we obtain

$$dE = 2\pi \rho_* \left(1 - \kappa + \frac{\kappa}{g_1}\sqrt{g_1 - g^2}\right) dg \tag{7.42}$$

and by integration

$$E = 2\pi \rho_* \left[(1-\kappa) g_1 + \frac{\pi \kappa}{4} g_1\right]. \tag{7.43}$$

In the absence of the deflecting body [Eq. (1.13)]

$$e = \pi g_1^2 \left(1 - \frac{\kappa}{3}\right) \tag{7.44}$$

and the ratio

$$\frac{E}{e} = \frac{2\rho_*}{g_1} \frac{12 - 2.58\kappa}{12 - 4\kappa}. \tag{7.45}$$

The actual case lies within the limits $0 < \kappa < 1$ or for ratio

$$\frac{2\rho_*}{g_1} < \frac{E}{e} < 1.18 \frac{2\rho_*}{g_1} \tag{7.46}$$

As the darkening coefficient grows towards the violet end of the spectrum, the occulted star becomes a little bluer than outside the occultation.

For great value $g > \frac{\rho_*}{5}$ the above approximation is not permitted.

For the disk with constant surface brightness ($\kappa = 0$) we have

$$\frac{E}{e} = \frac{\sqrt{g_1^2 + 4\rho_*^2}}{g_1}. \tag{7.47}$$

This formula can be derived from the image of the occulted star. According to our findins [Eq. (7.60)] the image is a ring of radii $\rho_{1,2}$ computed from Eq. (7.53) with $g = g_1$. The solid angle covered by it will be

$$\frac{\pi(\rho_1^2 - \rho_2^2)}{l_2^2} \tag{7.48}$$

and in the absence of the deflecting star

$$\pi \frac{g_1^2}{l_2^2}. \tag{7.49}$$

The ratio of both expressions leads again to the expression [Eq. (7.47)].

7.6. Normalised Gravitational Occultation

Our equations for dioptrical and photometrical aspects of the occultation can be normalised by the introduction of the critical distance Eq. (7.16) as the unity for the angles and we get then

$$\rho_{1,2} = \tfrac{1}{2}[g \pm \sqrt{g^2 + 4}], \tag{7.50}$$

$$I_{1,2} = \frac{g \pm \sqrt{g^2 + 4}}{4g\sqrt{g^2 + 4}}, \tag{7.51}$$

$$I = \frac{g+2}{g\sqrt{g^2+4}} \approx \frac{1}{g} + 0{,}37g. \tag{7.52}$$

The Table 7.2 gives several numerical values of the above quantities and the Table 7.3 the intensity of illumination produced by the stellar disk of the given normalised angular radius g_1 at some values of the angular distances γ between the centres of the deflecting an occulted body. The ratio $g_1/16'$ is given at the bottom of Table 7.3. The values of the illumination are given in decadic logarithms for three darkening degrees, namely $\kappa = 0$; $\kappa = 0{,}5$; $\kappa = 1{,}0$.

Table 7.2 [4]

g	ρ_1	ρ_2	I_1	I_2	I
0.0	1.0000	1.0000	—	—	—
0.05	1.0253	0.9753	10.5092	9.5092	20.0184
0.1	1.0512	0.9512	5.5195	4.5195	10.0390
0.2	1.1050	0.9050	3.0373	2.0373	5.0746

Table 7.2 (Continued)

g	ρ_1	ρ_2	I_1	I_2	I
0.3	1.1612	0.8612	2.2225	1.2225	3.4450
0.4	1.2198	0.8198	1.8238	0.8238	2.6476
0.5	1.2808	0.7808	1.5914	0.5914	2.1828
0.6	1.3440	0.7440	1.4417	0.4418	1.8835
0.7	1.4095	0.7095	1.3394	0.3394	1.6788
0.8	1.4770	0.6770	1.2659	0.2659	1.5318
0.9	1.5466	0.6466	1.2118	0.2118	1.4236
1.0	1.6180	0.6180	1.1708	0.1708	1.3416
1.1	1.6912	0.5912	1.1392	0.1392	1.2784
1.2	1.7662	0.5662	1.1145	0.1145	1.2290
1.3	1.8427	0.5427	1.0950	0.0950	1.1900
1.4	1.9206	0.5206	1.0793	0.0793	1.1586
1.5	2.0000	0.5000	1.0667	0.0667	1.1334
1.6	2.0806	0.4806	1.0564	0.0564	1.1128
1.7	2.1624	0.4624	1.0479	0.0479	1.0958
1.8	2.2454	0.4454	1.0410	0.0410	1.0820
1.9	2.3293	0.4293	1.0352	0.0352	1.0704
2.0	2.4142	0.4142	1.0303	0.0303	1.0606
2.1	2.5000	0.4000	1.0263	0.0263	1.0526
2.2	2.5866	0.3866	1.0228	0.0228	1.0456
2.3	2.6740	0.3740	1.0200	0.0200	1.0400
2.4	2.7620	0.3620	1.0174	0.0175	1.0349
2.5	2.8508	0.3508	1.0154	0.0154	1.0308
2.6	2.9401	0.3401	1.0136	0.0136	1.0272
2.7	3.0300	0.3300	1.0120	0.0120	1.0240
2.8	3.1204	0.3204	1.0106	0.0106	1.0212
2.9	3.2114	0.3114	1.0095	0.0095	1.0190
3.0	3.3028	0.3028	1.0085	0.0085	1.0170
3.1	3.3946	0.2946	1.0076	0.0076	1.0152
3.2	3.4868	0.2868	1.0068	0.0068	1.0136
3.3	3.5794	0.2794	1.0061	0.0061	1.0122
3.4	3.6723	0.2723	1.0055	0.0055	1.0110
3.5	3.7656	0.2656	1.0050	0.0050	1.0100
3.6	3.8592	0.2592	1.0046	0.0045	1.0091
3.7	3.9530	0.2530	1.0041	0.0041	1.0082
3.8	4.0471	0.2471	1.0037	0.0037	1.0074
3.9	4.1414	0.2414	1.0034	0.0034	1.0068
4.0	4.2360	0.2360	1.0031	0.0031	1.0062
4.1	4.3309	0.2309	1.0028	0.0028	1.0056
4.2	4.4260	0.2260	1.0026	0.0026	1.0052
4.3	4.5212	0.2212	1.0024	0.0024	1.0048
4.4	4.6166	0.2166	1.0022	0.0022	1.0044
4.5	4.7122	0.2122	1.0020	0.0020	1.0040
4.6	4.8080	0.2080	1.0019	0.0019	1.0038
4.7	4.9039	0.2039	1.0017	0.0017	1.0034
4.8	5.0000	0.2000	1.0016	0.0016	1.0032
4.9	5.0962	0.1962	1.0015	0.0015	1.0030
5.0	5.1926	0.1926	1.0014	0.0014	1.0028

Table 7.3 [4]

$16'\gamma$	g_1					
g_1	0.001	0.005	0.01	0.05	0.1	0.3
	3.3005	2.6016	2.3005	1.6017	1.3010	0.8280
0'	3.3336	2.6330	2.3319	1.6331	1.3324	0.8585
	3.3757	2.6767	2.3757	1.6767	1.3758	0.9020
	3.2909	2.5920	2.2909	1.5922	1.2911	0.8195
5'	3.3156	2.6166	2.3156	1.6166	1.3158	0.8432
	3.3503	2.6514	2.3503	1.6515	1.3506	0.8768
	3.2539	2.5549	2.2539	1.5551	1.2545	0.7832
10'	3.2632	2.5642	2.2621	1.5643	1.2639	0.7931
	3.2767	2.5777	2.2767	1.5780	1.2772	0.8055
	2.9466	2.2476	1.9466	1.2480	0.9489	0.4900
20'	2.9430	2.2440	1.9430	1.2445	0.9450	0.4843
	2.9376	2.2386	1.9376	1.2393	0.9395	0.4814
	2.7436	2.0437	1.7437	1.0461	0.7490	0.3118
30'	2.7425	2.0426	1.7426	1.0445	0.7482	0.3118
	2.7399	2.0400	1.7400	1.0426	0.7443	0.3118
	2.3746	1.6758	1.3747	0.6821	0.4014	0.0792
68'	2.3744	1.6755	1.3746	0.6830	0.4031	0.0792
	2.3741	1.6752	1.3742	0.6821	0.4014	0.0755
	2.2068	1.5080	1.2074	0.5237	0.2648	0.0334
100'	2.2066	1.5078	1.2071	0.5224	0.2624	0.0294
	2.2063	1.5074	1.2068	0.5224	0.2624	0.0294
$10^7 g_1/16'$	625	3125	6250	31250	62250	187500

7.7. Images of the Occulted Star

With the aid of the principles explained in 7.2, we are able to draw the images of the star during its gravitational occultation. The suitable formula is Eq. (7.14) from which we eliminate the angle r by Eq. (7.13)

$$\rho_{1,2} = \frac{1}{2}\left(g \pm \sqrt{g^2 + \frac{4Kk\alpha_2^2}{\alpha_1 + \alpha_2}}\right) \quad (7.53)$$

and substitute there from Eq. (7.16) the critical distance ρ_*

$$\rho_{1,2} = \frac{1}{2}\left(g \pm \sqrt{g^2 + 4\rho_*^2}\right). \quad (7.54)$$

Our Table 7.2 may equally be used for this purpose.

For the construction of the images we use the principles of 7.2 as shown in Fig. 7.9. In the first phase of the occultation we see two lenticular features situated outside and inside the critical circle ρ_*.

17 Link, Eclipse Phenomena

258 Einstein's Deflection of Light

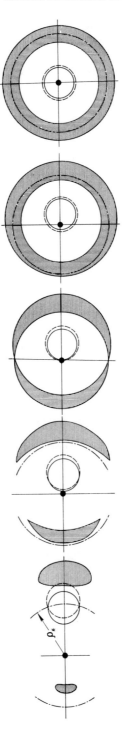

Fig. 7.9. Imaging action of Einstein's deflection

Images of the Occulted Star

As the limb of occulted star is approaching the centre of the deflecting star, the shape of both images takes the form of two crescents whose cusps come into contact at the moment when the limb touches the centre.

In this particular phase the contours of both crescents are two circles (Fig. 7.9 in the middle). One, the circle with the radius ρ_* and its centre at D (Fig. 7.10), is evidently the image of the element whose $r=0$. For

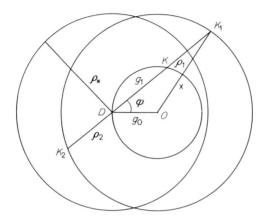

Fig. 7.10. Circular form of images

the other, we can show that any point K_1 at its circumference has from the centre 0 of the occulted star the same distance x. From Fig. 7.9 we find easily for K

$$g = 2g_1 \cos \varphi \tag{7.55}$$

and for the distance of its image K_1

$$\rho_1 = g_1 \cos \varphi + \sqrt{g_1^2 \cos^2 \varphi + 1}. \tag{7.56}$$

The distance from 0 will be therefore

$$x^2 = \rho_1^2 + g_1^2 - 2\rho_1 g_1 \cos \varphi \tag{7.57}$$

or after substitution from Eq. (7.56)

$$x^2 = 1 + g_1^2. \tag{7.58}$$

The case of circular contours arrives also at the central occultation when the image has the form of a ring whose width is [Eq. (7.50)]

$$\Delta \rho = \rho_1 + \rho_2 = g_1 \tag{7.59}$$

and the mean radius

$$\rho_m = \frac{1}{2}(\rho_1 - \rho_2) = \rho_* \sqrt{1 + \frac{g_1^2}{4\rho_*}} \approx \rho_* \left(1 + \frac{g_1^2}{8\rho_*}\right). \tag{7.60}$$

17*

Most of the dioptric phenomena display in the neighbourhood of the critical circle ρ_*. It is therefore necessary to examine the visibility of this region from the Earth with regard to the angular radius α_2 of the deflecting star. We have to examine the maximum possible value of the ratio

$$\frac{\rho_*}{\alpha_2} = \sqrt{\frac{Kk}{\alpha_1 + \alpha_2}}. \tag{7.61}$$

When the occulted star is in the infinity ($\alpha_1 = 0$) we obtain

$$\frac{\rho_*}{\alpha_2} = \sqrt{\frac{Kk}{\alpha_2}} = \sqrt{\frac{4}{3}\pi k \frac{a_\odot}{m_\odot}} \sqrt{a \delta l_2} \tag{7.62}$$

where δ is the density of the deflecting star and l_2 its distance from the Earth. This ratio is most favourable if the deflecting body is big (a), dense (δ) and distant (l_2). Unfortunately enough, big stars are usually thin and small ones are dense.

7.8. Tichov's Investigations [5]

TICHOV published two papers on the problem one 1937 and another in 1938. We shall analyse here the second and more detailed paper. This will be done in our transcription of symbols and in a simplified presentation. TICHOV replaces the observer N in the plane II (Fig. 7.11) by the object glass of diameter $2s$ receiving the light from the point source at M. He considers two beams issued from M, the geometric going directly from M to N and the first beam deflected by the star D.

If the geometric ray describes the circumference of the object glass, the deflected ray in its prolongation as far as the plane II describes there a kind of ellipse having the semiaxes s_t and s_r. The ratio of solid angles of both beams gives evidently the modification of the illumination by the deflecting body. We have for this ratio

$$\frac{I}{i} = \frac{\text{ellipse}}{\text{circle}} = \frac{s_t s_r}{s^2}. \tag{7.63}$$

TICHOV computes separately

$$R_{ad} = \frac{s_r}{s} \quad \text{the radial ratio} \tag{7.64}$$

$$T_{an} = \frac{s_t}{s} \quad \text{the tangential ratio}. \tag{7.65}$$

For the radial ratio we have evidently

$$R_{ad} = \frac{d\rho}{dg} = \frac{d\rho}{dr} \frac{dr}{dg} \qquad (7.66)$$

or with aid of Eqs. (7.14)–(7.15)

$$R_{ad} = \frac{1}{2}\left(1 + \frac{r}{\sqrt{r^2 + 4(\alpha_1 + \alpha_2)Kk}}\right). \qquad (7.67)$$

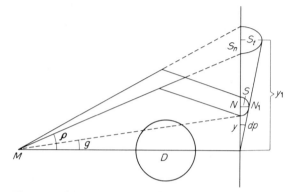

Fig. 7.11. Tichov's treatment of the photometrical action

In order to understand better the tangential ratio let us turn the ray MN around the axis MD so that the point N comes to the periphery N_1 of the object glass covering the radius s. At the same time the prolongation of the deflected ray covers the path s_t. We have hence

$$dp = \frac{s}{y} = \frac{s_t}{y_1} \qquad (7.68)$$

where dp is the necessary angle of rotation around the axis MD. Thus we have

$$T_{an} = \frac{y_1}{y} = \frac{\rho}{g} \qquad (7.69)$$

or with Eqs. (7.14) and (7.15)

$$T_{an} = \frac{1}{2}\left(1 + \frac{\sqrt{r^2 + 4(\alpha_1 + \alpha_2)Kk}}{r}\right). \qquad (7.70)$$

The product

$$R_{ad} \times T_{an} = \left(\frac{I}{i}\right)_1 \qquad (7.71)$$

gives the modification of the illumination for one beam which is identical with our Eq. (7.34). TICHOV calls it the ratio of total luminosities and gives a similar equation for the second conjugate beam.

On the axis for $g=0$ the separate computation of R and T fails and the author deals with the ratio of solid angles of the object glass viewed both directly and after the deflection. He finds finally the formula

$$I_0 = \frac{1}{s}\sqrt{\frac{\alpha_2}{\alpha_1}\left(\frac{\alpha_2}{\alpha_1}+1\right)}\, a\,K\,k \qquad (7.72)$$

which passes after some simplifications into Eq. (7.35). Here both components I_1 and I_2 are included.

A peculiarity of Tichov's paper is that the major part of it is based on the Newtonian deflection ($\frac{1}{2}$ of that of EINSTEIN), including the numerical examples, and only at the end (2 pages out of 19) does he introduce Einstein's deflection with corresponding equations as the contribution of his young pupil BOGORODSKYI.

7.9. Refsdal's Investigation

REFSDAL [3] in his treatement of our problem seems to be inspired by Tichov's first paper on this theme. However, not having found it easy to follow — the present author shares entirely Refsdal's opinion — he carries on in a clear geometrical way. For the illumination by the point source, he derives the expression identical to Eq. (7.34). Likewise the expression for the total illumination

$$I = \frac{1}{2}\left(\frac{\rho_1-\rho_2}{g}+\frac{g}{\rho_1-\rho_2}\right) \qquad (7.73)$$

is identical with (7.35). In addition REFSDAL states that

$$I_1 - I_2 = i$$

and further

$$\frac{I_1}{I_2} = \frac{R_1^2}{R_2^2} = \frac{\rho_1^2}{\rho_2^2}. \qquad (7.74)$$

Finally Refsdal's approximations near the axis for I/i and E/e are identical with our corresponding Eqs. (7.41) and (7.46).

7.10. Liebes' Investigations [6]

LIEBES considers the dioptric and photometric aspects of Einstein's deflection. Investigating the formation of the image, he derives the quadratic equation for apparent directions

$$\rho^2 - g\rho - \rho_*^2 = 0 \qquad (7.75)$$

whose solution is identical with Eq. (7.14). Making use of the critical distance ρ_*, LIEBES calls it the cone of inversion.

He does not investigate the photometric case of the point source but attacks directly although approximately the problem of illumination by the stellar disk. He does it in three stages. For great distances from the axis, he assimilates the two images of the occulted star to "bent ellipses" and derives their solid angle proportional to the modification of the intensity. His expression

$$I_{1,2} = \frac{1}{4}\left(\pm 2 + \frac{\sqrt{g^2 + 4\rho_*^2}}{g} + \frac{g}{\sqrt{g^2 + 4\rho_*^2}}\right) \qquad (7.76)$$

is identical with our Eq. (7.34) for the point source illumination. In other words he makes the illumination by the stellar disk equal to the point source illumination at its center.

Nearer the axis when $\rho^* > g > g_1$, LIEBES gives without demonstration the approximation

$$\frac{E}{e} = \frac{\rho^*}{g_1} \qquad (7.41)$$

and for the central occultation according to the solid angle of the annular image, he writes the expression

$$\frac{E}{e} = \frac{2\rho^*}{g_1} \qquad (7.45)$$

identical with our formula [Eq. (7.45)] for small values of g_1 and $\kappa = 0$.

In the two Liebes statements: that the angular thickness of the ring is very nearly equal to g_1 and its mean angular radius equal to ρ^* the adverbial phrase "very nearly" should be shifted from the first to the second statement.

7.11. Consequences of Einstein's Deflection in the Stellar Universe

Although the investigations which have just been explained are clear enough and concordant with the expected phenomena, their occurrence in the stellar universe must be examined with care from both theoretical and observational view-points.

The necessary condition for deflection phenomena is, of course, a nearly perfect alignment of three interested bodies. Firstly, the frequency and the duration in different stellar configurations, such as our

Galaxy, star clusters or other galaxies, must be examined and if possible computed or simply estimated.

Besides the subsidiary conditions concerning the visibility in relation to the dimension of the deflecting body, we must take into account the observability of the phenomenon in respect to its luminosity. In other words the intensification of the occulted star must be high enough to give an observable light amplitude for the whole system of occulted + deflecting bodies. Meanwhile, we renounce the detection of any change in shape or size due to the imaging action of the deflecting body.

Having found favourable circumstances for frequency, duration and light amplitude, we must discuss the observation side of the phenomenon. We must also bear in mind that the ephemeris is not generally available to focus our attention on a definite passage, so we have to rely on mere chance to find it between countless star images on the photographic plate. We have a very efficient method for the detection of any light variation of photographic images (blink microscope) but among the tremendous number of stars we have other serious candidates as variable stars. The question is how to decide in a given case of stellar variability between its intrinsic origin or the action of the deflecting body. Unfortunately, our terrestrial observations are largely limited in time by Earth's rotation and climatic conditions. Perhaps the proposed lunar or spatial observatory might guarantee better chances for the monitoring of any chosen stellar field.

LIEBES [6] computes or estimates the frequency and duration of gravitational passages as the function of the amplitude of light flash in the global intensity of occulted and deflecting stars. As this kind of estimation lies beyond the scope of this book, we shall give here only the basic assumptions and the numerical results. Liebes considers the following stellar configurations in their simple model structure:

   i) Our Galaxy observed in the direction of the Milky Way.

   ii) Deflecting star before a typical globular cluster.

   iii) Typical globular cluster as deflecting body before a star field in the direction of the galactic center.

   iv) Events within a single globular cluster.

For these configurations Liebes gives for several values of light intensification the frequency of the event and its duration measured between points of half intensity.

Adopting as the observable value $A_M \approx 2$ he found that the most favourable situation presents in the Milky Way (i) and then within a single globular cluster (iv). Nevertheless, we must wait some years in order to observe an event lasting only a few days. As for other configurations, unless a very unexpected chance arises the situation is practically hopeless.

In globular clusters we may, however, expect a statistical effect of Einstein deflection [1]. Towards the center of typical configuration the clustering of stars is great enough to produce an observable effect. Adopting the same mass and radius for two cluster stars as for the Sun and their mean distance D we obtain for the central occultation [Eq. (7.46)]

$$\frac{E}{e} = 38.8 \sqrt{D} \quad (D \text{ in parsecs}). \tag{7.77}$$

With some parsecs of distance between the stars the intensification can be observable. In fact we observe in globular clusters that their surface brightness increases towards the center more rapidly than would be expected from the distribution of stars within the cluster. This was formerly explained by the predominance of bright stars in the central regions of the cluster but the above interpretation also holds good.

Apologising for some unavoidable omissions, we shall quote briefly the following papers: HOAG [7] pointed out the existence of a particular object (AR 15 h 15.0 min; Decl. $+21°$ 36'; 1950) consisting of a perfect halo of 17" of radius around a central diffuse image. The gravitational effect may be invoked and the mass of the central object should be according to HOAG about 8×10^{11} of solar mass.

The paper of IDLIS and GRIDNEVA [8] examines only the dioptrical part of Einstein's deflection in the case of mutual occultation of two galaxies. BOGORODSKYI [9] considers the influence of the stars of our Galaxy on the luminosity of other distant galaxies. In the same context may be classified the paper of KLIMOV [10].

PAGE investigated [11] a very particular manifestation of Einstein's photometrical effect, namely the gravitational focusing of radiation behind a dense mass in an H-II region. In this case a bright spike should appear and it is found to be barely possible that this phenomenon may reveal intergalactic hydrogen or collapsed galaxies.

7.12. Gravitational Passage or Occultation

The case of the near passage of two stars with pronounced Einstein's deflection has been examined by REFSDAL [3] in order to determine the mass of deflecting star. Assuming with the author the constant spatial velocities of the three bodies involved, the angular geometric distance g of two stars will depend on the time t according to the simple formula

$$g(t) = \sqrt{g_0^2 + \mu^2 t^2} \tag{7.78}$$

where g_0 is the minimum distance at the time $t=0$ and μ the angular velocity of one star relative to the other. During the passage only the sum of the luminosities of both stars

$$L(t) = L_D + L_O(t) \tag{7.79}$$

can be measured and we have, therefore, with Eq. (7.41) the following basic equation

$$L(t) = L_D + L_O \frac{\rho^*}{\sqrt{g_0 + \mu^2 t^2}}. \tag{7.80}$$

This equation contains five unknown quantities namely L_D, L_O, μ and ρ^*, g_0. REFSDAL assumes that the first three can be determined before or after the passage when the two stars are optically separated. It remains to determine ρ^* and g_0. This can be done from the measurements of $L(t)$ at the moment $t=0$ and at another moment t during the passage. From the critical distance we find according to Eq. (7.16) the mass M if we know the distances of both stars from the Earth.

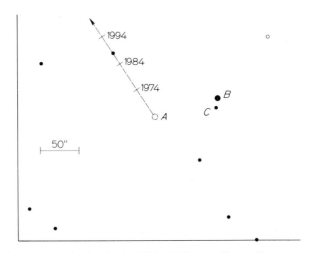

Fig. 7.12. Occultation by 40 Eridani (A) according to FEIBELMAN

FEIBELMAN reports [12] on the possibility of actually observing such an occultation. Triple star 40 Eridani (5^m) will pass in 1988 across the 15^m star (Fig. 7.12). An amplification of $1000 \times$ would be detectable from the measurements and with the amplification $10000 \times$ the phenomenon could be observed without difficulty. The parameters of 40 Eridani being known, a quantitative verification of the theory will be possible.

7.13. History of Einstein's Photometrical Effect

LIEBES in his above quoted paper [6] has reviewed the history of Einstein's photometrical effect. We shall use it with some necessary supplements.

This history can be traced back to 1919 when LODGE [13] protested against the use of the expression "gravitational lens" noting that it would be "impermissible to say that the solar gravitational field acts as a lens, for it has no focal length". In 1920 EDDINGTON [14] had been wrong noting only the weakening action of the gravitational deflection which is, however, only a component of the total effect. It seems according to ZWICKY [17], that FROST as early as 1923 outlined a program for the search for Einstein's effect among the stars.

In 1924 CHWOLSON published a brief paper [15] where he expressed the idea that Einstein's deflection of light in the gravitational field of a star can give an image of another remote star. In the case of central occultation the image has an annular form. The paper is, however, without any mathematical formula and the photometrical consequences are not mentioned. Chwolson's paper remained unknown until 1937 when TICHOV, his neighbour in Pulkova, quoted it in his paper [5].

Curiously enough, the idea of photometric consequences of Einstein's deflection emerged almost simultaneously and independently during the year 1936. For us (March) it was a natural extension of our formula giving the attenuation of light by refraction [Eq. (1.35)] to the gravitational deflection [1]. For EINSTEIN (December) it was the answer to his correspondent MANDEL who suggested the possibility of a gravitational lens [16]. In our first paper [1] we obtained the Eq. (7.20) for the general photometrical effect and the expression Eq. (7.16) for the critical distance. The occurrence of the effect in galaxies is briefly discussed. EINSTEIN [16] gives the formula for the critical distance [Eq. (7.16)] and for the total increase of the illumination [Eq. (7.35)]. The author is sceptical about the usefulness of the phenomenon in astronomical practice.

ZWICKY [17] in 1937 offers nothing new in the theory but suggests the possibility of the mutual occultation of two galaxies in order to determine the mass of the occulting galaxy. He finds an acceptable probability for such an occultation. In the same year we gave an complete and exhaustive theory of the phenomenon.

TICHOV, knowing Chwolson's paper and all the above mentioned work and having quoted them in his articles [5], has given an independent derivation of the formulae with identical results. Therefore, it is difficult to justify the conclusion of the DARWIN lecture by MICHAILOV [18] indicating TICHOV as the author of the photometric theory of Einstein's deflection without quoting his predecessors.

Bibliography

1. LINK, F.: Compt. rend. **202**, 917 (1936); — Bull. Astr. **10**, 73 (1937).
2. EDDINGTON, A.: Mathematische Grundlagen der Relativitätstheorie. Braunschweig 1932.
3. REFSDAL, S.: M.N. **128**, 295 (1964); **134**, 315 (1966).
4. LINK, F.: BAC **18**, 215 (1967).
5. TICHOV, G.A.: C.R. Acad. Sci. Moscou **16**, 199 (1937); — Bull. Pulkovo **16**, No. 130 (1938).
6. LIEBES, S., JR.: Phys. Rev. B **133**, 835 (1964).
7. HOAG, A.A.: Astr. J. **55**, 378 (1950).
8. IDLIS, G.M., i S.A. GRIDNEVA: Izv. Astrophys. Inst. Ak. Nauk Kazak. S.S.R. **9**, 78 (1960).
9. BOGORODSKYI, A.F.: Publ. Kiev Astr. Obs. **11** (1962).
10. KLIMOV, J.G.: Doklady Akad. Nauk S.S.S.R. **148**, 789 (1963).
11. PAGE, T.L.: Smithsonien Inst. Astrophy. Obs. Spec. Rep. 196 (1965).
12. FEIBELMAN, C.: Science **151**, 73 (1966).
13. LODGE, O.J.: Nature **104**, 354 (1919).
14. EDDINGTON, A.S.: Space, time and gravitation, p. 134, 308. New York 1920.
15. CHWOLSON, O.: A.N. **221**, 329 (1924).
16. EINSTEIN, A.: Science **84**, 506 (1936).
17. ZWICKY, F.: Phys. Rev. **51**, 290, 679 (1937).
18. MICHAILOV, A.A.: M.N. **119**, 608 (1959).

Subject Index

Aerosol layer on Jupiter   201
− − on Mars   192, 196
− −, terrestrial   40, 66, 69
− − on Venus   221, 223
Air mass   33
− −, climatic influences   36
− −, mean   29
− − theory, classical   51
− − −, modern   33
Atmosphere on Jupiter   190, 201
− on Mars   186, 196, 237
− on the Moon   230
−, terrestrial   35, 38
− on Venus   188, 214, 242

Bessel's refraction theory   52
Bouguer's graph   22, 67

Chapman's ionospheric model   228, 239
Critical distance   233, 247, 257
− height   184, 243

Danjon's cat's eye photometer   56
− luminosity scale   97
− relation   97, 100
Density gradient   34, 35, 36, 188
Dioptrics of dense atmosphere   182
− of thin atmosphere   180

Eclipses of Jovian satellites   197, 200, 202
− of Phobos   193
− of space probes by Mars   193
− − − − by Venus   193
Einstein's deflection, general   244
− −, Chwolson's work   267
− −, dioptrics   246
− −, history of the photometrical effect   267
− −, imaging action   257
− −, Liebes' work   262
− −, normalised occultation   255

Einstein's deflection, path difference   252
− −, photometry   248, 250, 253
− −, Refsdal's work   262
− − in stellar universe   263
− −, Tichov's work   260
− −, Zwicky's work   267
Extinction coefficient, astronomical   21
− −, terrestrial   22
− −, theoretical   20

Fesenkov's refraction theory   53
Fraunhofer central, line intensity   96

Gravitational image   257

High absorbing layer   40, 66, 71, 176
Hipparchos' formula   104
Hot spots on the Moon   119

Invariance relation of refraction   182
Ivory's refraction theory   52

Laplace's air mass theory   51
Linke's diagram   70
Lomosonov's phenomenon   215
Luminescence, lunar   88, 92, 96
−, −, outside eclipses   94
−, −, in penumbra   90
−, −, recent work   96
−, −, in umbra   99
Lunar eclipses   2
− −, atmospheric illumination   42
− −, brightness   97
− −, catalogues   9
− −, central   2
− − in chronology   11
− − in cislunar space   46
− −, ephemeris   3, 8, 59
− −, future   8
− −, geographical circumstances   4
− −, geometrical conditions   1
− −, global intensity   84

Lunar eclipses, history 9, 11
– –, magnitude 5, 8
– – and meteoric activity 77, 112
– –, meteorological perturbations 78
– – at microwaves 120
– – on the Moon 47, 85
– –, partial 2, 8
– –, penumbral 2, 9, 91
– – and solar activity 97, 100
– –, total 2, 8
– –, tropospheric influences 77
– –, volcanic influences 81
– surface, theoretical model 118
– –, thermal inertia 116

Mariner-IV occultation 236
Mariner-V occultation 241
Meteoritic accretion 75
– particles 71
– –, fall in the atmosphere 71
Millikan's regime of fall 72
Moon, global intensity 94
–, – –, fluctuations 94, 95

Occultation, gravitational 255
–, –, predicted 266
– radio 227
– – by Mars 236
– – by the Moon 228, 230
– – by solar corona 231
– – by Venus 241
Occultations, optical 180
–, –, far 184
–, –, by Jupiter 189
–, –, by Mars 186, 192
–, –, near 192
–, –, terrestrial 203
–, –, by Venus 187, 192
Opposition effect 92
Oriani's theorem 51
Ozone, atmospheric 62
–, absorption coefficients 62
–, amount 40, 65, 145
–, density 90
–, distribution with height 41, 64, 145, 149, 151
–, distribution with latitude 63, 65

Penumbra, light excess 91
–, photoelectric measurements 91
–, problematic variations 102
–, semidiameter 1

Penumbra, theory, complet 89
–, –, simplified 88
Penumbral eclipses, future 8

Rayleigh's law 20
Refraction, climatic, variations 36
–, confrontation with observations 35
– in dense atmosphere 182
–, electronic 226
–, image 48, 86, 142, 206, 220
– on Jupiter 190
– light attenuation 24, 129, 190
– on Mars 192
– modification of the light intensity 228, 229, 232
–, optical 33
– theory, classical 51
– –, modern 33
– in thin atmosphere 180
–, terrestrial 33
– on Venus 192, 205, 213

Satellite, actives 121, 152
–, passives 121
– eclipses 122
– –, work at Arcetri 157
– –, – at China Lake 147
– – of Echo 2 146, 150
– –, ephemeris 124
– –, Fesenkov's theory 144
– –, geographical circumstances 141
– – in L-alpha 154
– –, observed from the satellite 122, 141
– –, work at Ondřejov 147, 150
– –, photoelectric photometry 146
– –, secondary illuminations 143
– –, work at Slough 160
– – of SR series 152
– – of SR-I 154
– – of SR-IV 157, 160
– – of SR-VII 157
– –, theory 128
– –, –, simplified 131
– –, work at Valensole 151
– –, visibility 123, 127
– – in X-rays 157, 160
Scattering of light on aerosols 19, 21
–, molecular 19
Scale height 181
– – on Jupiter 192
– – on Mars 192, 237

Subject Index

Scale height, Nicolet's formula 161
— — on Venus 188, 192, 242
Shadow, auxiliary 27
—, auxiliary isophotes 78
—, climatic variations 38
—, density definition 15
—, —, normal 28
— increase 104
— —, explanations 110
— —, Hartmann's method 106
— —, history 104
— —, Kosik's method 106
— —, Maedler's method 105
— — and meteoric activity 112
— —, Paetzold's discussion 111
— —, results 108
— —, Schilling's discussion 114
— —, Seeliger's simulation 111
— isophotes 79
— flattening 108, 113
— measurements 55
— —, photoelectrical 58
— —, photographic 57
— —, visual 55, 60
— terminator 6, 61, 141
—, theory, classical 49
—, —, modern 14
Solar disc integration 15, 128, 172
— limb darkening 15

Solar minima by lunar eclipses 100
Stoke's regime of fall 73

Thermal inertia of the Moon 116
— radiation of the Moon 115
— — during the eclipses 116, 118
Transmission coefficient general 19, 129
Twilight azimuth profiles 178
— balloon measurements 175
— in cislunar space 46
— components 70, 169
— correction of night light 170
—, fundamental problems 174
— illumination of the upper atmosphere 171
—, light excess 176

Umbra photometrical theory 14
— semidiameter 1

Venus aureole 209
— cusps extension 216
— — —, Dollfus' work 222
— — —, Edson's work 221
— — —, Russell's explanation 221
— — —, Schoenberg's work 220
— polar caps 210, 213
— rotation 213
— transits over the Sun 205, 209

Type-setting, printing and bookbinding: Universitätsdruckerei H. Stürtz AG, Würzburg